SUFFOLK UNIVERSITY
MILDRED F. SAWYER LIBRARY
8 ASHBURTON PLACE
BOSTON, MA 02108

D1227789

ABOUT VISUAL STRATEGIES

Visual Strategies is an essential guide to communicating scientific findings and concepts through graphics for researchers in all disciplines.

Any scientist or engineer who communicates research results will immediately recognize this practical handbook as an indispensable tool. The guide sets out clear strategies and offers abundant examples to assist researchers—even those with no previous design training—with creating effective visual graphics for use in multiple contexts, including journal submissions, grant proposals, conference posters, and presentations.

Visual communicator Felice Frankel and systems biologist Angela DePace, along with experts in various fields, demonstrate how small changes can vastly improve the success of a graphic image. They dissect individual graphics, show why some work while others don't, and suggest specific improvements. The book includes analyses of graphics that have appeared in such journals as *Science, Nature, Annual Reviews, Cell, PNAS,* and the *New England Journal of Medicine,* as well as an insightful personal conversation with designer Stefan Sagmeister and narratives by prominent researchers and animators.

ABOUT THE AUTHORS

Felice C. Frankel is a research scientist in the Center for Materials Science and Engineering at MIT and the recipient of numerous awards and honors for her work in visual communication. Among her previous books is *Envisioning Science: The Design and Craft of the Science Image.* Angela H. DePace is an assistant professor in the Department of Systems Biology at Harvard Medical School, where her lab studies the mechanism and evolution of gene regulation. They both live in Boston. Stefan Sagmeister, a leading graphic designer and typographer, has a design firm in New York City.

ADVANCE PRAISE

"In this technoscientific century, with knowledge doubling every decade, researchers and designers alike need to ramp up their presentation of the material they describe. This beautifully illustrated book shows how."
—*Edward O. Wilson, University Research Professor Emeritus and Honorary Curator in Entomology, Harvard University*

"A thoughtful and useful series of recommendations that will actually help you understand what you are doing when you are trying to make yourself clear."
—*Milton Glaser*

"Scientists presenting even simple data to busy journal readers are well advised to invest some thought in their visual comprehensibility and impact. This unique book provides exactly what they need: copious case studies across the disciplines, wise principles and the authors' outstanding creativity, experience and integrity—in both technical and ethical senses—in visualizing the results of science."
—*Philip Campbell, Editor-in-Chief,* Nature

"Anyone—scientist or not—who is interested in using pictures to teach, to convey information, or to catch attention must study this book. It is splendid. In it you learn: what information can be conveyed graphically, how to design images for maximum intelligibility and interest, how to draw in the reader, and what successful images look like. As a bonus, you get a cheerfully readable style, you learn about some extremely interesting research, you see how some very good researchers, drawn from across science, think about what they do in terms of images, and you have the pleasure of a brilliantly laid-out book."
—*George M. Whitesides, Woodford L. and Ann A. Flowers University Professor, Harvard University*

"Unique ... an essential guide to literacy for fields that are essential to all our lives."
—*Steven Heller, School of Visual Arts*

VISUAL
STRATEGIES

Felice C. Frankel & Angela H. DePace

VISUAL STRATEGIES

A PRACTICAL GUIDE TO GRAPHICS
FOR SCIENTISTS & ENGINEERS

Yale University Press / New Haven and London
Design by Sagmeister Inc.

Copyright ©2012 by Felice C. Frankel and Angela H. DePace.

All rights reserved.

This book may not be reproduced, in whole or in part, including illustrations, in any form (beyond that copying permitted by Sections 107 and 108 of the US Copyright Law and except by reviewers for the public press), without written permission from the publishers.

Yale University Press books may be purchased in quantity for educational, business, or promotional use. For information, please e-mail sales.press@yale.edu (US office) or sales@yaleup.co.uk (UK office).

Design by Sagmeister, Inc.
with CHIPS

Set in Foundry Gridnik and Adobe Garamond Pro.

Printed in China.

ISBN 978-0-300-17644-5

Library of Congress Control Number: 2011937511

This paper meets the requirements of ANSI/NISO Z39.48-1992 (Permanence of Paper).

10 9 8 7 6 5 4 3 2 1

OVERVIEW

WHY THIS GUIDE?

Graphics—visual representations of scientific data and concepts—are critical components of science and engineering research. Images engage us in ways that words cannot. Explanatory graphics can clarify or strengthen an argument by guiding us through data or concepts. Exploratory graphics draw us into the research process, allowing us to discover patterns and relationships ourselves.

Visual representations have long been a significant part of any scientist's and engineer's research. Until fairly recently it was standard practice for universities and research institutions to hire specialists to help researchers visually communicate their work. Now the research community is primarily responsible for crafting its own graphics—and yet the typical researcher's training rarely includes the development of such skills and sensibilities.

This guide will help you create more effective graphics to support your work. Together, we will explore specific examples from journal articles, presentations, grant submissions, and other scientific contexts. We will make practical suggestions, based on our work with science and engineering researchers, to help you answer the question, "What is the *best graphic* to communicate my work?"

It is important to remember that a visual representation of a scientific concept (or data) is a re-presentation, and not the thing itself—some interpretation or translation is always involved. There are many parallels between creating a graphic and writing an article. First, you must carefully plan what to "say," and in what order you will "say it." Then you must make judgments to determine a hierarchy of information—what must be included and what could be left out? The process of making a visual representation requires you to clarify your thinking and improve your ability to communicate with others. Furthermore, the process of making an effective graphic often leads to new insights into your work; when you make decisions about how to depict your data and underlying concepts, you must often clarify your basic assumptions.

What we provide in the pages of this guide is an approach: a set of strategies that are a distillation of what we have learned, both independently and in collaborative projects, over many years. We believe these strategies will help you create improved representations of your work. These strategies might even make you reconsider your discipline's classic visualizations and ask yourself, "How effectively do the standard approaches communicate?" Whether you are inventing a new representation or updating a classic, we encourage you to invest the time to create a good graphic. The best ones have the power to shape new thinking and approaches in your field.

SPEAKING OF DESIGN: A CONVERSATION

We were privileged to work with designer Stefan Sagmeister for this guide. In many ways, good design is the kernel that began our thinking for this project. In addition to seeing his design on these pages, we wanted to bring you his voice.

FCF, AHD The fact that you decided to design our book is pretty remarkable for us, Stefan. Is science something you were always interested in?

SKS No, not at all, in fact I had little interest in science when I was in school, received average grades, and took science education for the most part as a necessary evil. I only developed an interest after going to TED and hearing from all these amazing people talking about all these amazing developments.

FCF, AHD We have always believed there has been a desperate need for talented graphic designers to participate in the process of communicating scientific data and concepts. Our readers might not immediately see the connection between your world and theirs. We wonder if you can help them think about it.

SKS During the last decade scientific research has made fantastic progress, but this has often been poorly communicated. This is true for communications between various scientific fields, science communications within peer reviewed media, as well as how science is treated in mass media.

FCF, AHD When you first saw some of the figures we were going to include, the "befores," did you see a visual thread of some sort in terms of the mistakes most researchers make?

SKS I was surprised to see a lot of very basic design mistakes. It seems many scientists creating complex visualizations are not aware of simple rules regarding color, form, or hierarchy. In numerous cases all it would take to fix this would be a design 101 course.

FCF, AHD That would be a fantastic idea, but realistically, in our experience, most researchers wouldn't give it the time. So, what would you think about the possibility of engaging more designers in the process? For example, do you think design schools might offer a specialized course for scientific graphics?

SKS Yes, I am convinced there is a tremendous interest among design students, specifically at the master's programs. Many are disillusioned with working for the commercial sector and would love the opportunity to sharpen their design skills dealing with content that engages their minds to the fullest.

FCF, AHD Do you think the design community would embrace the idea of working on material with which they are not familiar? Do you think it would be important for designers to understand the material?

SKS Yes and yes. Designers work with unfamiliar material all the time; learning about a new sector, business, or scientific process is a very integral part of the job (and I'd say one part that makes that job rather enjoyable). And yes, they had better understand the material; otherwise, innovative thinking or focused visualization will be impossible.

FCF, AHD And specifically addressing *Visual Strategies*, if you remember our meeting on that hot NYC day on your office terrace, your request to place a subtle 5% yellow background behind all the images was a concern for the two of us. We were worried, as most scientists tend to be, that we were changing the data in a way—that we were augmenting the original content. However, you made the case that the design of the book warranted the change. We are sure our readers will have their own opinions about that matter. They will have the opportunity to discuss the issue with us on our online *Visual Strategies* forum [see Appendix]. Can you clarify how a change like that benefits the design, and, more important, in general, what is your advice on design changes to a figure in science?

SKS This is a wonderful question: whether a visualization should be more "objective" or "subjective" comes up in various ways on almost every single job. Are we communicating more effectively by placing more value on the overall form or are we better off sticking to the process in a literal way? Will more people understand if we use a sexy visual conceit or are we better off sticking to established modernism?

Does the colorization undermine or promote understanding? In the case of the cream backgrounds within this book, I myself don't see a problem at all. They simply are a more orderly way of organizing the graphics on the page, a more sophisticated alternative to the black hairline.

FCF, AHD Thanks so much, Stefan. The process of working with you has inspired us to think differently about creating scientific graphics. Let's hope that more scientists and engineers will have the opportunity to closely collaborate with designers.

Acknowledgments

In 2003, after the two of us serendipitously met at a Gordon Research Conference (GRC) on visualization in science and education, one of us wrote to the other, "One way to emphasize common ground and to begin to define a vocabulary of visual solutions would be to organize the discussion around categories of visual solutions rather than around the specific concepts being represented." And so, without knowing it at the time, we began work on this guide. We are grateful to the GRC for creating an environment where exciting collaborative ideas are encouraged to take root.

We first worked together on the Image and Meaning conferences and workshops at MIT (www.imageandmeaning.org). We extend our deepest thanks to the National Science Foundation for funding those national participatory workshops at which scores of researchers and graphic designers came together to investigate new approaches to the development of scientific graphics. We are grateful to all the Image and Meaning participants and workshop leaders, with a special thanks to Rebecca Perry and Rosalind Reid, who were also instrumental in the development of this guide.

Visual Strategies would have never been possible without the generosity of all the contributors whose exemplary work is displayed in the pages of this guide, along with their names and affiliations. We are indebted to them all for their patience and responsiveness.

We are grateful to the journals and researchers who granted us permission to use their figures for our discussions. We would like to specifically thank the illustrators at Annual Reviews (AR), including Doug Beckner, Glenda Mahoney, Fiona Martin, and Eliza Jewett-Hall, for not only sharing their own work, but also for diligently searching through the issues of AR for relevant graphics. Our deepest gratitude is extended to the editors of AR: Ike Burke, Veronica Padilla, and Jennifer Jongsma, for their enthusiasm and support for this project.

Thanks to the DePace lab and Systems Biology Department at Harvard Medical School for cheerfully providing critical feedback on the guide's usefulness for practicing scientists, and to Robin Heyden, Rachael Brady, and Rebecca Ward for their additional comments and guidance.

Our deepest thanks go to Karen Gulliver for her expertise in editing our sometimes-not-so-readable prose; to Teddy Blanks of CHIPS, a NYC-based design studio, for his outstanding efforts in implementing and augmenting Stefan's design; to Michelle Suave, who helped with initial layouts; and to Jean Thomson Black at Yale University Press for her years of experience and savvy in shepherding projects through all the systems.

And finally, creative work is never possible without the continued patience and support of our friends and family, who we simply cannot thank enough.

— Felice Frankel and Angela DePace, Boston 2011

A ROADMAP: HOW TO USE THIS GUIDE

GUIDE
ORGANIZATION

Form and Structure, Process and Time, Compare and Contrast
The first three chapters will help you define the purpose of your graphic. We identify three major types of scientific graphics: those that illustrate form and structure; those that illustrate processes over space and time; and those that encourage readers to compare and contrast. We explore examples of each of these types used in current research; we define the purpose of the graphic, suggest improvements, and present a revised version. You might note that some of the examples could comfortably reside in more than one chapter, which points to interesting overlaps in concepts and principles.

Case Studies
In this chapter, we explore selected works by researchers and designers in depth, written in their own words. Along with the "before" and "after" figures, the text describes the process and includes the "in-betweens," with explanations of their decisions. The stories describe the evolution of selected figures and animations—and how the researchers' thinking shaped the process.

Interactive Graphics
In this chapter we take a closer look at interactive graphics. Some are explanatory animations. Some are exploratory graphics, sometimes called data visualizations. In both cases, we describe what the interactivity brings to the graphic. The lessons learned about static graphics in the first four chapters will also apply to these interactive examples. The principles are the same. The strategies in this book are fundamental to any form of visual expression and will always provide a solid starting point for thinking about your visual expression, regardless of its final form.

Visual Index
Here you will find a grid of images relevant to each example in the guide and citations for each for quick visual reference.

Appendix
The Appendix contains a brief description of the website associated with this book and suggested further readings.

The design of this guide reflects the way in which we hope you will use it: as a source of inspiration and as a workbook to refine your critical thinking skills regarding graphics. We present a great variety of examples because we think that the best way to learn how to make effective graphics is by dissecting why some graphics are effective and others are not—and determining how they can be improved. We also present the strategies used to create each graphic in a succinct visual format, to allow you to quickly see how these strategies are applied in multiple contexts.

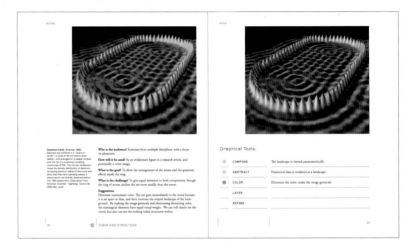

Show by Example

Throughout these pages, we explore different types of graphics to illustrate how they were imagined, created, and refined. We encourage you to consider all of the examples—static images and animations—because it is our belief that the challenges in making good graphics are consistent across different disciplines and all types of representations. Many examples are displayed on two-page spreads to demonstrate how a figure can be improved, represented by a "before" (on the left) and an "after" (on the right) version. A caption provides a brief description of the image and its reference (see the example above). We offer brief answers to the following questions: *Who is the audience?* For whom is your figure intended? How will it be used? In what format will the graphic appear, e.g., a journal article, textbook, grant submission, presentation? *What is the goal?* What do you want the viewer to see? *What is the challenge?* What is difficult about achieving that goal?

We then offer suggestions for improving the graphic—suggestions made either by us or by the researcher or designer credited in the caption. There is, of course, not a single correct way to visually represent data or concepts. In fact, you might disagree with how we arrived at a particular "after" representation—or you may have additional ideas for improving an illustration. We welcome those conversations and invite you to share them on our website forum (see Appendix). We hope that these examples will encourage debate and help you develop an analytical approach to creating graphics—just as scientists do in their research.

Tabs

The colored tabs on the right edge of the pages identify the chapters of the guide. We included tabs to visually separate the different sections, and to allow you to easily find a section that interests you. *Form and Structure, Process and Time*, and *Compare and Contrast*, marked with orange tabs, identify simple concepts that can be tackled in preparing graphics. In *Case Studies* (with a yellow-orange tab) and *Interactive Graphics* (a chartreuse tab), these concepts are explored in more depth.

Graphical Tools:

●	COMPOSE	The landscape is viewed axonometrically.
●	ABSTRACT	Numerical data are rendered as a landscape.
●	COLOR	Eliminate the color; make the image grayscale.
	LAYER	
	REFINE	

Graphical Tools Grid

On the bottom right of each two-page spread you will find a grid to help you quickly identify which graphical tools were used in each example to achieve the goals of the adjoining graphic. Dots indicate which tools are used in the figure: Gray dots indicate tools that were used in the original graphic, and remain unchanged. Orange dots indicate tools that were used to improve the figure.

YOUR FIRST STEP: ASK YOURSELF BEFORE YOU BEGIN

1. Is the graphic *explanatory* or *exploratory* (perhaps even interactive)?

Explanatory graphics are used to communicate a point or call attention to patterns and concepts. Explanatory graphics can be used as evidence or proof in research, and can be teaching tools for colleagues and students. They can also be powerful teaching tools for you, the person creating them. The process of planning a graphic requires you to first clarify the primary point of the graphic—what the figure should include. Explanatory graphics can also be interactive, such as animated images that the viewer can play, pause, and reverse. Static explanatory graphics are considered in *Form and Structure*, *Process and Time*, *Compare and Contrast*, and *Case Studies*. Interactive explanatory graphics are considered in *Interactive Graphics*.

Exploratory graphics (sometimes referred to as visualizations) invite the viewer to discover information. Many scientific disciplines generate enormous datasets. New graphical approaches are required to make sense of the data and to organize and communicate the main points. Interactive exploratory graphics are also considered in *Interactive Graphics*. All good graphics, whether they are explanatory or exploratory, are based on the same set of principles.

2. How will the graphic be used?

Scientific graphics are used in many contexts: in oral presentations to colleagues, in lectures, as part of research articles, in posters, and in grant proposals—to name just a few. Each of these contexts makes different demands on a graphic. For example, in an oral presentation, a graphic needs to make the point quickly and clearly because an audience does not have the time to contemplate a graphic in nearly as much depth as it might in a research article. Graphics may have to appeal to different types of audiences as well. A graphic that is intended for a colleague may be quite different from a graphic intended for a student in your course, and still different for a program officer or a congressperson without a background in your field. Your answer to this important question will help shape the content and form of the figure.

3. What is the first thing you want the viewer to see?

While you might know exactly how to navigate your explanatory graphic to find the crucial information, it is unlikely that the first-time viewer will see exactly what you see. Without your experience and familiarity with the topic, the audience needs to be guided to identify important details in the sea of information. A viewer's eye must be guided to "read" the elements in a logical order. The design of an exploratory graphic needs to allow for the additional component of discovery—guiding the viewer to first understand the overall concept and then engage her to further explore the supporting information.

YOUR
TOOLS

We present some basic tools from graphic design that you can use to enhance the clarity of your science graphics. These tools are Compose, Abstract, Color, Layer, and Refine. Here we explain each tool and present two examples in which that tool has been used successfully.

● COMPOSE   **Organize the elements and establish their relationships.**

● ABSTRACT   **Define and represent the essential qualities and/or meaning of the material.**

● COLOR   **Choose colors to draw attention, to label, to show relationships (compare and contrast), or to indicate a visual scale of measure.**

● LAYER   **Add layers to overlap multiple variables to create a direct relationship in physical space.**

● REFINE   **Edit and simplify.**

Organize the elements and establish their relationships.

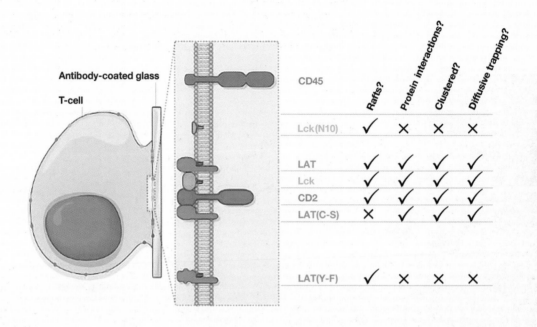

Spatial organization of signaling proteins.
Nature, 2005.
Single-molecule tracking reveals that spatial organization of signaling proteins in the cell membrane, often ascribed to lipid-based "rafts," is probably a consequence of protein-protein interactions.

Composition helps the audience understand how to "read" the graphic—where to start and where to move on. Notice how this figure is arranged. Its composition creates a visual hierarchy that indicates a narrative meant to be read, intuitively, from left to right. Attention to the composition allows you to immediately see that the illustration comprises three sections, and to understand the relationships among those sections. At the far left we see a T-cell. The use of clear composition and consistent color-coding makes clear that the second section is a detail of the plasma membrane, where lipids and important proteins are abstracted using color and familiar forms. Color-coding and proper placement in the third section provides information about each protein. With the composition flowing from left to right, a sweep of the eye establishes the narrative of each protein, inviting us to look closely for detail and pattern.

In contrast to our first graphic, a composition of a typical left-to-right organization, the figure below is organized around a central image. Each image is shown in relation to the overall structure of the tree. Detailed structures of the leaves, trunk, and roots are shown at the left. The position of the forest on a map is shown at the right. Note how the composition clarifies the interrelationships among all five images.

Plant-water interactions. *Annu Rev Ecol Evol Syst*, 2007.
Water moves from the soil to the atmosphere through roots, xylem, and stomata.
a The root-soil system is heterogeneous at a wide range of scales (smallest ~0.1 μm).
b The xylem within the individual branches generates a complex network (scale ~100 μm).
c Plant-atmosphere gas exchange is controlled by stomata (scale ~10 μm).
d Turbulent eddies transport water vapor from stomata to the free atmosphere.
e Atmospheric states are modulated by the landscape heterogeneity (scale ~10 km).

● ABSTRACT

Define and represent the essential qualities and/or meaning of the material.

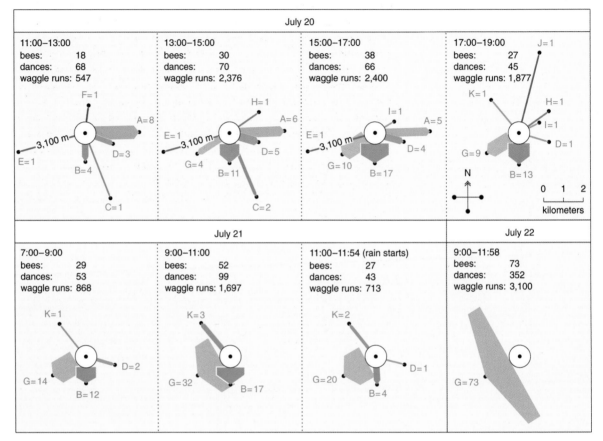

July 20

| 11:00–13:00 | | 13:00–15:00 | | 15:00–17:00 | | 17:00–19:00 |

11:00–13:00
bees: 18
dances: 68
waggle runs: 547

13:00–15:00
bees: 30
dances: 70
waggle runs: 2,376

15:00–17:00
bees: 38
dances: 66
waggle runs: 2,400

17:00–19:00
bees: 27
dances: 45
waggle runs: 1,877

July 21   **July 22**

7:00–9:00
bees: 29
dances: 53
waggle runs: 868

9:00–11:00
bees: 52
dances: 99
waggle runs: 1,697

11:00–11:54 (rain starts)
bees: 27
dances: 43
waggle runs: 713

9:00–11:58
bees: 73
dances: 352
waggle runs: 3,100

History of a bee swarm decision-making process. *Behav Ecol Sociobiol*, 1999. The circle within each of the panels represents the location of the swarm; each colored bar pointing out from the circle indicates the distance and direction of a potential nest site. The number of different bees that danced for the site in the time period is indicated by the number at the tip, and the width of the bar is proportional to the number. The total number of bees, dances, and waggle runs summed over all the potential nest sites for the time period shown is listed at the top.

When you decide how to depict your data, you decide on the abstraction. Will you present a graph? A cartoon? An accurate molecular model? And which features will you include in these representations? Your preferred abstraction should include all necessary information, exclude unnecessary information, and make use of your reader's preexisting knowledge without being confined by it. It's a tall order, but fulfilled by these two examples.

Consider the illustration above, depicting bee swarm behavior. The original data were presented as large, hard-to-read tables of numbers. In this imaginative re-creation, the authors use abstract shapes to represent both the population of bees and the direction in which they travel. Without knowing the precise quantities, the reader can grasp the conclusions through the size, color, and shape of the representations.

In the example below, the challenge was to represent the function, rather than the physical structure, of DNA molecules in an assembly process. To introduce the new abstraction, the author borrows from a familiar representation and relates the two. The top row depicts the binding of DNA molecules; text labels and connecting lines indicate complementary base-pairs. A new abstraction is introduced in the bottom row; only the inputs and outputs are shown. Note that this composition allows the viewer to learn the new abstraction by comparing steps with a more familiar representation. This new abstraction is successful for a number of reasons. Most importantly, it is compact; the old representation becomes unwieldy when trying to show the assembly of many molecules. Moreover, it presents a new way of thinking about how to design molecules to assemble in a predictable way. First, the researcher can determine how to coordinate all of the inputs and outputs, and then decide how to design the necessary base-pairing relationships. The graphic thus depicts an important change in thinking and provides a comprehensible visual representation of a particular reaction.

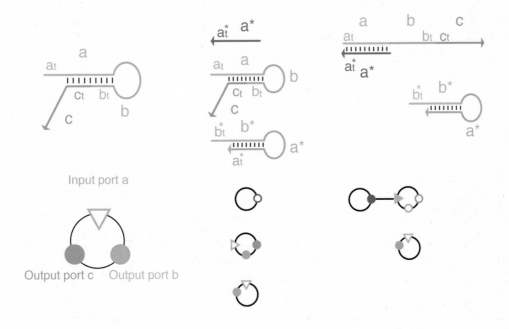

Dynamic DNA assembly. Presentation slide, 2010.
The assembly of DNA hairpins can be designed and controlled. (Top) Secondary structure mechanism illustrating an assembly reaction during catalytic duplex formation. Asterisks denote complementarity. (Bottom) Abstraction of the same reaction using a node motif with three ports. Triangles represent inputs; circles represent outputs. Open circles are accessible ports; filled circles are inaccessible ports. Colors are consistent with the top panel to allow comparison.

Choose colors to draw attention, to label, to show relationships (compare and contrast), or to indicate a visual scale of measure.

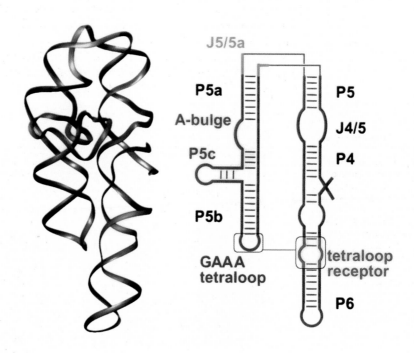

RNA structure. *Annu Rev Phys Chem*, 2001. A schematic for secondary structure of the P4-P6 domain of the *Tetrahymena* ribozyme. The base-paired (P) stems are labeled. On the left a ribbon diagram of the crystal structure of the P4-P6 domain, in the same orientation as the secondary structure, is shown. Comparison of the two gives the locations of the tertiary interactions between the base-paired stems.

Color can tell us where to look, what to compare and contrast, and it can give us a visual scale of measure. Because color can be so effective, it is often used for multiple purposes in the same graphic—which can create graphics that are dazzling but difficult to interpret. Separating the roles that color can play makes it easier to apply color specifically for encouraging different kinds of visual thinking. Do you want to show a reader where something is? Would you like to encourage comparison between two components, or be able to follow components through a process? Perhaps you have a large dataset and need to separate different elements visually. Or perhaps you would like to enable your reader to analyze quantitative data visually. These two examples use color to label and to provide quantitative information.

In the figure above, the reader can connect areas on the 3D structural representation on the left to those same areas in the schematic of base-pairing on the right, simply using color.

In the figure below, color is used to indicate numerical value, in this case, distance. The challenge these authors faced was to show the relative position of equivalent cells in embryos of two related species of fruit flies. Because embryos are three-dimensional, one way to see all of the cells at once is to "unroll" the embryo using a cylindrical projection. The authors encoded positional information in this unrolled view in two ways. First, each pair of cells is indicated by a line that extends from the cell in one species to the cell in the other, where it terminates with a white dot. The dots give the lines directionality and the entire graphic a sense of motion. Second, because the unrolled representation of the embryo results in significant spatial distortion, the length of the line doesn't correspond to the distance between two cells. Instead, the authors apply a color-code: blue for near and red for far.

Relative position of cells in two embryos.
Presentation, 2010.

For each cell in one species, the distance and direction to the average position of the top 10 best corresponding cells in another species are shown. The correspondence is shown with a line that starts at the position of the query cell, and ends at the average position of the target cells. The end of the line is indicated with a white dot. Because the 2D projection distorts actual distance in 3D, the lines are color-coded to indicate actual distance traversed in 3D. Blue is a short distance; red is a longer distance.

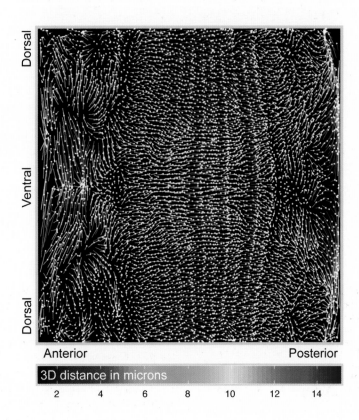

Add layers to overlap multiple variables to create a direct relationship in physical space.

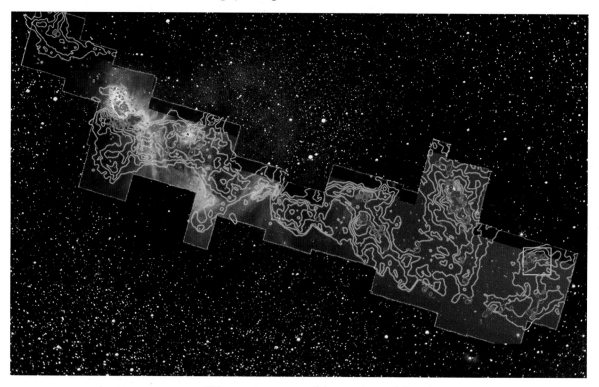

Perseus. Online resource, 2011.
Multiple views of the Perseus molecular cloud complex, overlaid to demonstrate the locations of different physical environments, and the spatial relationships between them. There are three layers: a background image of stars, a false-color, infrared layer showing heat (red is warm, blue is cool), and finally contours showing the location of clouds of molecular carbon monoxide. These clouds are depicted using red/blue ellipses showing dust emissions at wavelengths of 1.1 mm and 0.85 mm, respectively, and correspond to the positions of the densest portions of the cloud. The yellow inset box indicates a region of interest that is explained in another figure not presented here.

When various types of data are layered directly on top of one another, the viewer is able to spatially correlate multiple features. This is immediately intuitive in the case of spatial relationships, such as in the example above. The Perseus molecular cloud complex is a vast cloud of dust and gas about 850 light-years from Earth. The cloud, which contains about ten thousand times the mass of our sun (or 4 billion times the mass of Earth), was formed out of the more diffuse gas which exists everywhere in the galaxy. Some patches of the cloud are denser than others and, in the densest pockets, gravity has already begun pulling together lumps of material to form new stars. The new stars unleash fast winds that sweep away remaining material and can either slow down or even channel the formation of further stars. Astronomical telescopes operating across visible, infrared, and radio wavelengths are required to observe the vast range of temperatures and densities of material in the cloud. A coherent picture emerges in which young stars appear to lie inside or near warm regions (in fact, the stars are warming up the regions around themselves), which in turn lie inside the densest patches of molecular gas.

This next figure uses layering to indicate changes in proportion over time at multiple scales. The authors are interested in how ecosystems change over long evolutionary timescales, in terms of the number and types of different species and how they interact with their environment. Graphs, therefore, show the total number of species over time during a given time period; each graph comprises multiple colored bands indicating how different species contribute to the total, giving the reader immediate insight into proportion. The figure also stacks multiple versions of these graphs along the vertical axis. This can be considered a different type of layering; though elements are not juxtaposed directly on top of one another, they are spatially related to one another along a single axis. Because each graph represents a set of species that dominated during a particular time period, comparing them this way shows the transitions between different sets of species over time.

Proportional diversity of species clusters through time.
Annu Rev Earth Planet Sci, 2010.
Each higher taxon within each cluster is colored according to its feeding mechanism, as indicated in the legend. Timescale abbreviations: €, Cambrian; O, Ordovician; S, Silurian; D, Devonian; C, Carboniferous; P, Permian; Tr, Triassic; J, Jurassic; K, Cretaceous; and Cen, Cenozoic. A subset of the data presented in the original figure is shown.

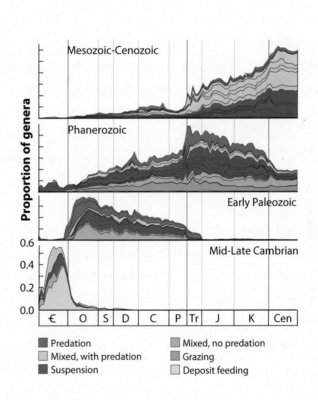

Edit and simplify.

BEFORE

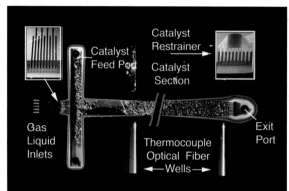

AFTER

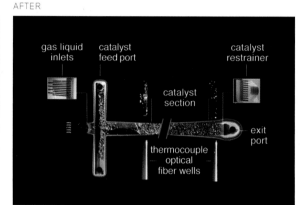

A microfabricated multiphase packed-bed reactor.
Ind Eng Chem Res, 2001.
A microchemical device built in silicon and glass by using microfabrication methods including deep-reactive-ion etch technology, photolithography, and multiple wafer bonding. The microchemical system consists of a microfluidic distribution manifold, a microchannel array, and a 25-μm microfilter for immobilizing solid particulate material within the reactor chip and carrying out different heterogeneous chemistries.

The final step in creating your graphic is to refine it. Step back and look at it with fresh eyes. Is there anything that could be removed? Or anything that should be removed because it is distracting? Consider each element in your figure and question whether it contributes enough to your overall goal to justify its contribution. Also consider whether there is anything that could be represented more clearly. Perhaps you have been so effective at simplifying your graphic that you could now include another point in the same figure. Another method of refinement is to check the placement and alignment of your labels. They should be unobtrusive and clearly indicate which object they refer to. Consistency in fonts and alignment of labels can make the difference between something that is easy and pleasant to read, and something that is cluttered and frustrating.

In this figure of a microfabricated device, the authors labeled components and inserted scanning electron micrographs (SEMs) to show detailed areas.

The figure above was edited and simplified in the following ways:
- Font size was reduced and the capital letters were changed to lower case.
- Unnecessary arrows were deleted.
- Positioning of the text was made more regular.
- SEM insets were re-oriented to match the direction of the larger image.

Another way to refine graphics is to combine data — but only if doing so clarifies rather than obscures the meaning of the image. For example, the original data for this figure were presented in two separate images — a colored graph showing absorption and fluorescence, and a photograph of the corresponding fluorescing vials. Combining these two types of data into a single image clarifies the correspondence and allows both types of data to be understood simultaneously.

BEFORE

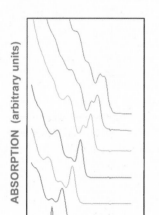

AFTER

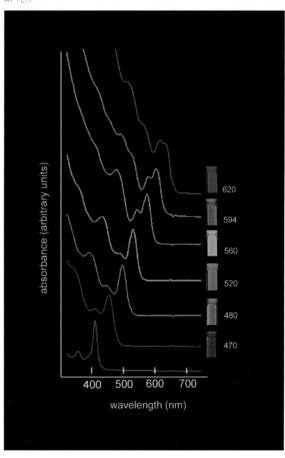

Nanocrystals. *J Phys Chem B,* 1997.
The wide spectral range of bright fluorescence from different size samples of (CdSe)ZnS contained in six vials is plotted along with their photoluminescence peaks occurring at 470, 480, 520, 560, 594, and 620 nm.

FORM AND STRUCTURE

There are multiple types of representations of structure and form in science. Some show physical structure directly, such as photographs or micrographs. Others show physical structure using a rendering of data, such as molecular models or topographical maps. These representations may be evidentiary (proving that something exists, or showing it in context), or seek to highlight a particular feature of the structure. We also consider many graphs to be representations of structure in a dataset. This may be less intuitive, but think about which features are often communicated in graphs, for instance, the shape of the distribution of values or the form of a function. Here, too, the goal is often to prove that something exists and/or, to describe its shape. In this first chapter you will see examples from astronomy, chemistry, physics, and biology.

A spectacular example of a rendered image is the reconstructed 3D image of a child's mummy shown at the right. This figure artfully depicts both the exterior of the mummy and the child's skull within using a cutaway approach. Reference slices showing various layers of the specimen are presented to the left. These two components fulfill the goals of the figure: to show the structure of the mummy and to tell the reader how the image was created. We made one change to the original figure—we converted the mummy's image to grayscale to differentiate between the skull and the mummy's encasing.

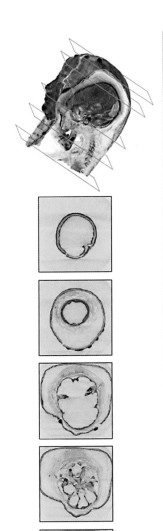

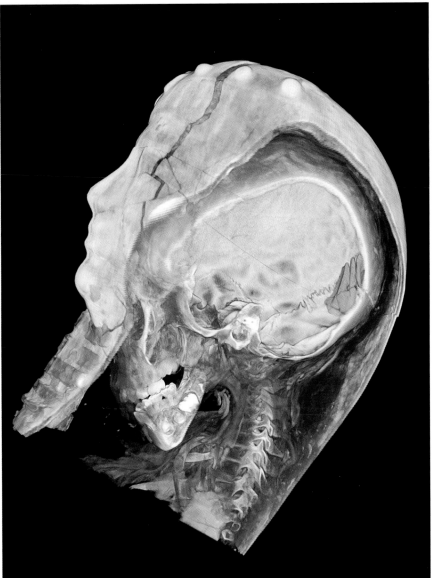

Mummy. *American Scientist,* 2006.
The image within the figure is part of a larger three-dimensional image reconstructed from 60,000 high-resolution two-dimensional scans. The team was led by endodontist W. Paul Brown of the Stanford-NASA National Biocomputation Center, working with physicist Rebecca Fahrig in the Department of Radiology and other colleagues at the Stanford University School of Medicine. The complete figure was designed by Barbara Aulicino.

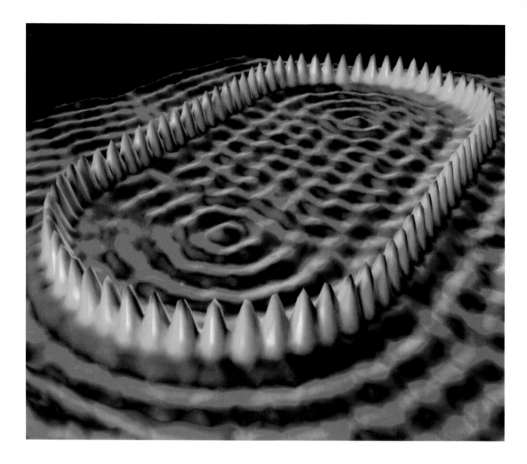

Quantum corral. *Science,* 1993.
Electrons are confined in a "quantum corral"—a circle of 48 iron atoms (blue peaks)—and arranged on a copper surface with the tip of a scanning tunneling microscope (STM). The circular oscillations reveal the density distribution of electrons occupying quantum states of the corral and show that they form standing waves, a phenomenon not directly observed before this 1993 experiment.

Who is the audience? Scientists from multiple disciplines, with a focus on physicists.

How will it be used? As an evidentiary figure in a research article, and potentially a cover image.

What is the goal? To show the arrangement of the atoms and the quantum effects inside the ring.

What is the challenge? To give equal attention to both components, though the ring of atoms catches the eye more readily than the waves.

Suggestions
Eliminate unnecessary color. The eye goes immediately to the corral because it is set apart in blue, and then traverses the striped landscape of the background. By making the image grayscale and eliminating distracting color, the topological elements have equal visual weight. We can still clearly see the corral, but we can also see the striking radial structures within.

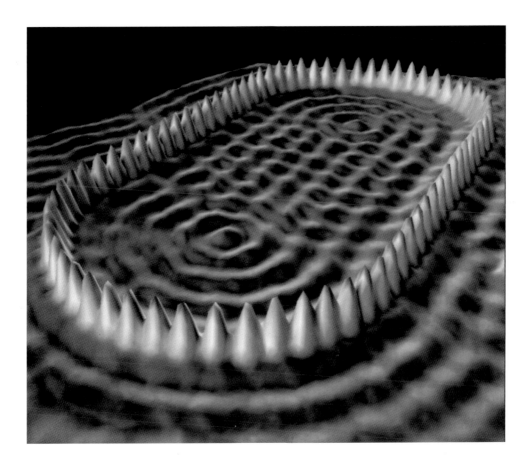

Graphical Tools:

●	COMPOSE	The landscape is viewed axonometrically.
●	ABSTRACT	Numerical data are rendered as a landscape.
●	COLOR	Eliminate the color; make the image grayscale.
	LAYER	
	REFINE	

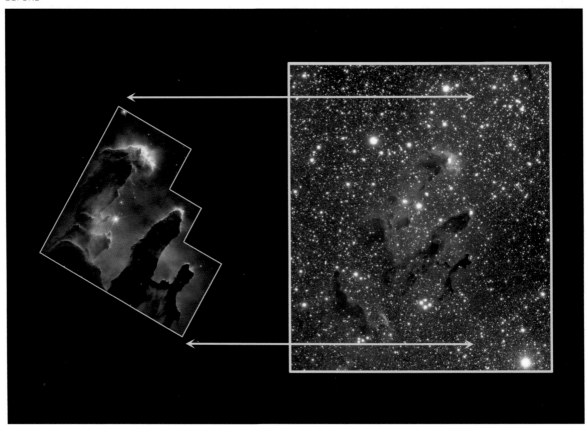

Eagle nebula. *Astronomy, from the Earth to the Universe*, 2002.
A comparison between the Hubble Space Telescope visible light image of the Eagle Nebula (called M16) and the same area of space imaged with a ground-based telescope reading the near-infrared part of the spectrum.

Who is the audience? Undergraduate students.

How will it be used? As a figure in an introductory textbook.

What is the goal? To illustrate the detailed structure of a nebula in context.

What is the challenge? To relate these two images to one another.

Suggestions
Clearly indicate the region of interest. In the original, the inset is ambiguous; there are similar structures visible in the image on the right, but the reader must closely scan a large area to find correspondences. Use an outline of the inset in a related color to indicate precisely where the inset is located in the larger image. Use a larger version of the context image to emphasize that the inset is a detail.

Graphical Tools:

●	COMPOSE	Use a larger photograph to provide context; display detail in an inset.
	ABSTRACT	
●	COLOR	Use a related color (green) to indicate the position of the inset.
●	LAYER	Place the outline of the inset directly on the background image.
●	REFINE	Remove arrows; because the inset shape is asymmetric and completely delineated, they are no longer needed.

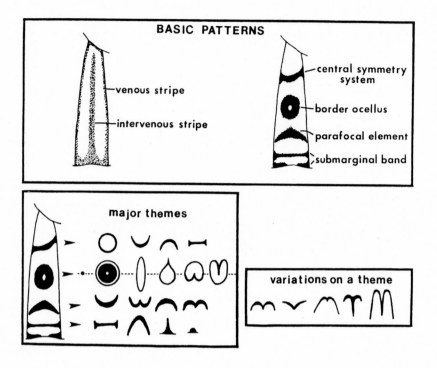

Butterfly wing patterns.
The development and evolution of butterfly wing patterns, 1991.
Major themes in butterfly wing cell patterns. Intervenous stripe patterns (top left) and nymphalid ground plan patterns (top right) form two mutually exclusive categories. Major themes for each pattern element (bottom left) and variations on a single theme (bottom right) are also illustrated.

Who is the audience? Advanced students and researchers.

How will it be used? As a figure in a specialty textbook.

What is the goal? To illustrate two main types of patterns and their variants.

What is the challenge? To indicate that patterns are related within two classes.

Suggestions

Consider a visual table. In the original, the data were presented in multiple panels that occurred in different sections of the book. By combining the data into a single figure, a hierarchy from major themes to variations can be shown directly. Delete unnecessary frames. The original panels were framed by bold outlines. These are unnecessary if the figure is recomposed.

FORM AND STRUCTURE

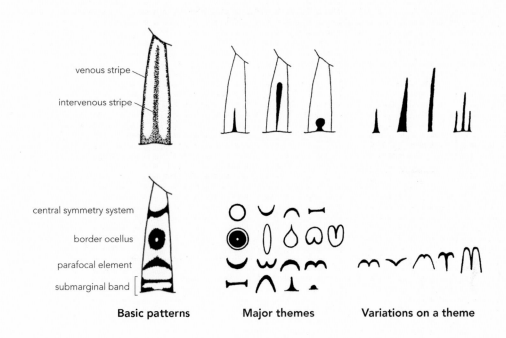

Basic patterns   **Major themes**   **Variations on a theme**

Graphical Tools:

●	**COMPOSE**	Arrange information from left to right—begin with major themes and relate them to relevant variations.
○	**ABSTRACT**	Patterns are represented with drawings.
	COLOR	
	LAYER	
●	**REFINE**	Remove unnecessary frames; add information where necessary; rearrange and standardize labels.

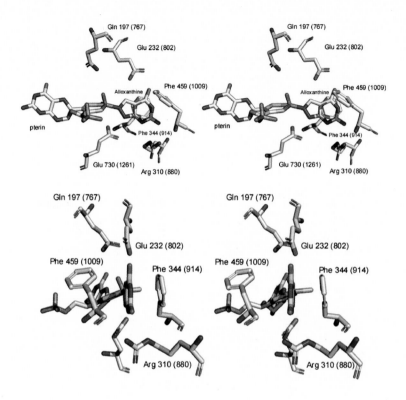

Enzyme structure. *J Biol Chem*, 2009. Two stereo pairs are shown, each depicting the active site of the enzyme xanthine dehydrogenase (entry 1jrp at the Protein Data Bank), including the pterin and molybdenum co-factors and an alloxanthine inhibitor, surrounded by the amino acids from the enzyme that interact with them. The two pairs are views from different angles; the bottom pair is a 90-degree clockwise rotation about the vertical axis of the top pair. Revision by David Goodsell.

Who is the audience? Scientists expert in molecular biology.

How will it be used? As a figure in a research article.

What is the goal? To show the interaction of protein side-chains with a substrate.

What is the challenge? To clarify 3D geometrical relationships in a 2D representation.

Suggestions

Do not show too much in a single image. The original sets of stereo images depict both hydrogen bonds and stacked amino acids. Using stereo images might not always be the best way of showing 3D relationships. An alternative is to split the image into two figures — one to show the hydrogen bonds (shown in the revised image) and another to present the interactions between amino acids. Focus attention using clear indicators — in this case, blue lines to show hydrogen bonds and different-sized balls to highlight different molecules.

FORM AND STRUCTURE

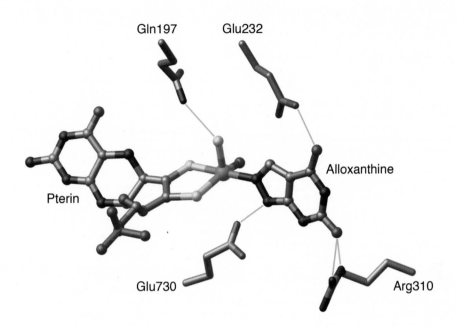

Gln197   Glu232

Alloxanthine

Pterin

Glu730   Arg310

Graphical Tools:

COMPOSE

●	ABSTRACT	Use different-sized balls to highlight different molecules; show hydrogen bonds with blue lines.
●	COLOR	Standard accepted colors are used.
●	LAYER	Do not layer if it inhibits clarity.
●	REFINE	Separate figures to increase clarity.

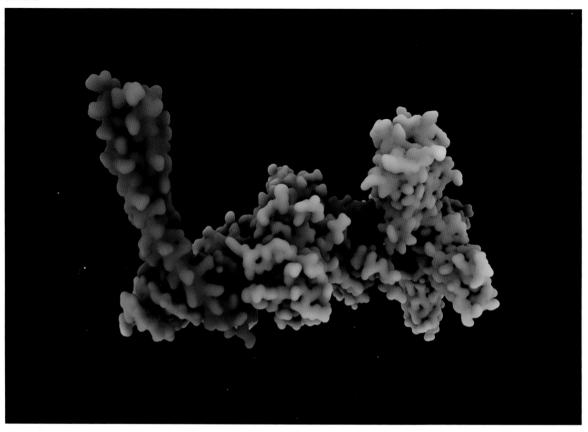

Molecular model. *Cover submission, 2007.* 3D molecular surface image of a large protein complex (the interferon beta enhanceosome) bound to DNA. The protein subunits are shown in red, orange, and yellow. DNA is shown in blue. All structural renderings are based on data from either X-ray diffraction or cryoEM. Revision by Gaël McGill.

Who is the audience? Scientists expert in biology.

How will it be used? As a cover submission for a journal.

What is the goal? To depict the complex's structure and function in the cell.

What is the challenge? To provide context without distracting from the details of the structure.

Suggestions

Present the subject in context. Place the complex in its cellular environment. Minimize the distraction of the surrounding environment by rendering it with a similar level of detail in faded blues. Use selective lighting to focus the viewer's attention on the relevant protein complex. Distinguish relevant proteins from the surrounding complexes with a different rendering: use a translucent shell and a colored ribbon diagram to provide both the overall shape and the details.

Graphical Tools:

●	COMPOSE	Place the complex of interest in a larger context.
●	ABSTRACT	Use multiple types of molecular renderings (surface only, translucent surface + ribbon diagram).
●	COLOR	Keep colors related: all DNA and cellular background are rendered in shades of blue; the enhanceosome is rendered in red, orange, and yellow.
	LAYER	
	REFINE	

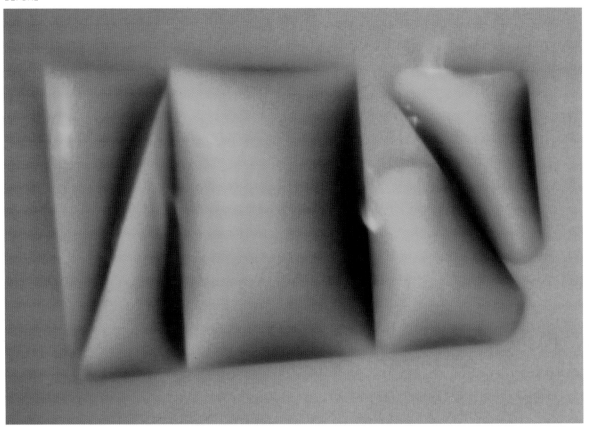

Square drops of water. *Science*, 1992.
A gold surface patterned with hydrophobic lines forming various shapes. The lines prevent water, dropped within the hydrophilic shapes, from spreading. The inner rectangle measures about 4 mm across.

Who is the audience? Researchers from multiple disciplines, with a focus on chemists.

How will it be used? As a figure in a research article, and potentially as a cover image.

What is the goal? To demonstrate that the patterning process creates both hydrophilic and hydrophobic areas.

What is the challenge? To indicate that the water droplets contained within the hydrophobic lines do not mix.

Suggestions
When possible, create a sample specifically designed to communicate the concept. In this case, adding dyes to the water dropped onto the defined areas creates an image that better communicates the fact that the hydrophobic lines prevent the water drops from mixing. A grid of more uniform shapes creates a more appealing image. An additional problem with the original image is that it is slightly out of focus.

 FORM AND STRUCTURE

Graphical Tools:

●	COMPOSE	Create a more uniform pattern that emphasizes the message.
	ABSTRACT	
●	COLOR	Color the two water samples before adding to the surface.
	LAYER	
●	REFINE	Photograph with the proper depth of field to keep everything in focus.

PROCESS AND TIME

Processes take place over time and result in change. However, we're often constrained to depict processes in static graphics, perhaps even a single image. Luckily, a good static graphic can be just as successful, perhaps even more so, than an animation. Giving the reader the ability to see each "frame" of time can offer a valuable perspective. In this chapter, we present static graphics depicting biological, material, and algorithmic processes.

The schematic illustration on the right shows transfer of solid printed objects from one surface to another. The authors chose the frames carefully, depicting only the critical steps. The degree of abstraction is appropriate for their goal of showing the overall transfer process. The researchers could have included considerably more detail about the chemistry of the surface, for example, but decided to keep it simple. This figure also uses color to distinguish steps. Even without reading the captions, we clearly see how the surface structures are lifted from one substrate and then "stamped" on to another. Finally, though the caption explains all of the steps, they are also succinctly described in the figure itself, making it easy for the reader to understand the figure on its own. For all of these reasons, this is a particularly successful example of a graphic that depicts a process.

A. Prepare donor substrate; apply rubber stamp

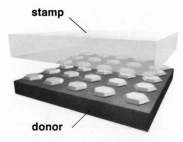

stamp

donor

B. Quickly peel back stamp; grab objects off of donor

C. Apply inked stamp to receiving substrate

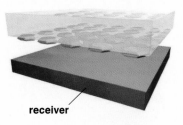

receiver

D. Slowly peel back stamp; print objects onto receiver

Transfer printing by kinetic control of adhesion to an elastomeric stamp. *Nature Materials*, 2006. The process begins with the preparation of microstructures on a donor substrate by solution casting, micromachining, self-assembly, or other suitable means. Laminating a stamp on the donor substrate and then quickly peeling it away pulls the microstructures from the donor substrate onto the stamp. Bringing the stamp in contact with another substrate and then slowly peeling it away transfers the microstructures from the stamp to the receiver. The peeling rate determines the strength of adhesion and, therefore, the direction of transfer.

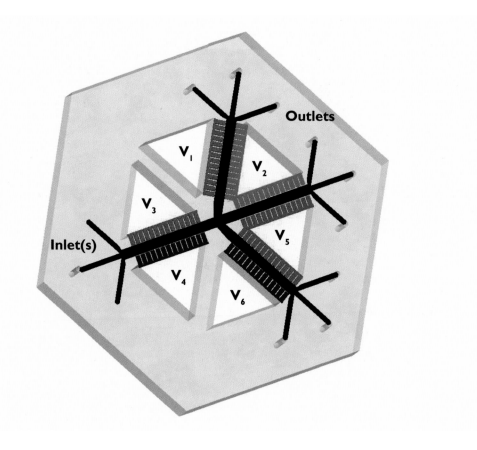

An electrophoretic separation device.
Presentation, 2006.
A mixture enters the inlet, flows through an
open channel (black), and is continuously
separated between gels with low pH (red)
and high pH (blue), resulting from different
electrical potentials (V_n). The first separation
stage is coarse; a finer separation occurs
between other pairs of gels with different pHs
(highest pH to lowest: blue>pink>green>red).

Who is the audience? Researchers and advanced students.

How will it be used? In a scientific presentation.

What is the goal? To show how material moves through the device and how
it is subject to different pH gradients.

What is the challenge? To follow both the movement and the conditions of
the fluid.

Suggestions
Use intuitive colors to help viewers with a scientific background understand
the color gradients within the gels, e.g., variations of blue are basic and
variations of red are acidic. Make paths explicit by adding arrows to indicate
the direction of the flow.

PROCESS AND TIME

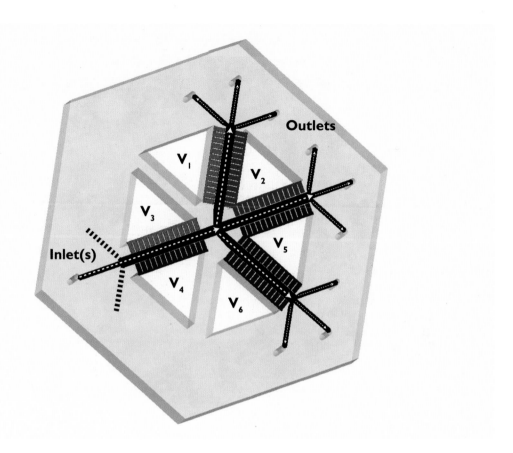

Graphical Tools:

●	COMPOSE	The flow of material is from left to right.
●	ABSTRACT	The device is shown as a 3D schematic.
●	COLOR	Use scientifically intuitive colors relating to pH.
●	LAYER	Add arrows on the inlet path to indicate direction of flow.
	REFINE	

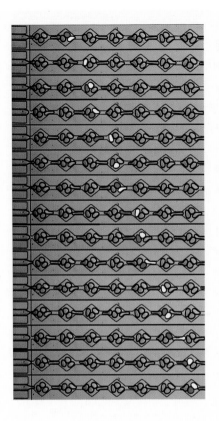

Bubbles moving in a microfluidic device.
Appl Phys Lett, 2004.
This figure shows a selection of "grab" shots from a video in which bubbles alternate between top and bottom positions as they flow through a channel. The addition of hand-coloring follows two selected bubbles.

Who is the audience? Researchers and advanced students.

How will it be used? As a figure for a research article.

What is the goal? To show the bubbles moving up and down relative to one another in a static image.

What is the challenge? To follow a particular set of bubbles over time.

Suggestions

Show only critical steps. There are many additional frames that could be included in this figure because they are from a movie. A limited number makes it easier to identify critical features. Clearly define your region of interest. Here, it is important to track a pair of bubbles to see how they move relative to each other. Choose easily distinguishable colors and highlight the background to show how the bubbles move over time.

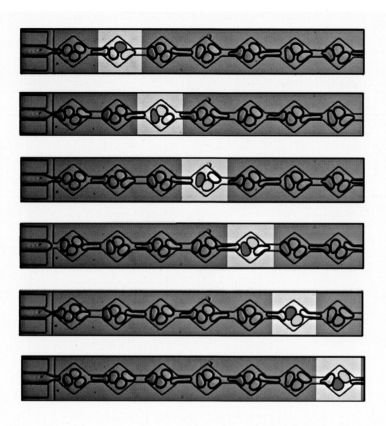

Graphical Tools:

●	COMPOSE	Time progresses from top to bottom; bubbles move left to right.
	ABSTRACT	
●	COLOR	A single pair of bubbles is red and yellow.
●	LAYER	Highlight the background behind the relevant pair of bubbles.
●	REFINE	Use fewer film grabs, just enough to make the point.

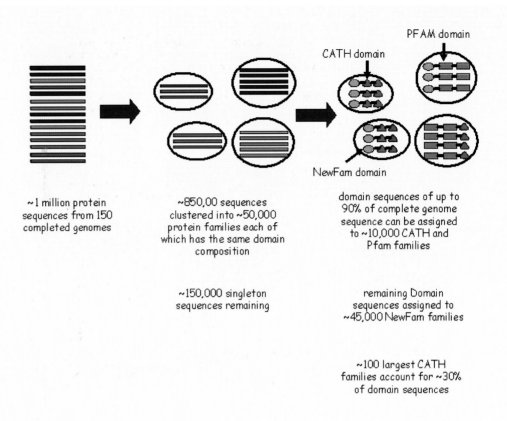

~1 million protein
sequences from 150
completed genomes

~850,00 sequences
clustered into ~50,000
protein families each of
which has the same domain
composition

domain sequences of up to
90% of complete genome
sequence can be assigned
to ~10,000 CATH and
Pfam families

~150,000 singleton
sequences remaining

remaining Domain
sequences assigned to
~45,000 NewFam families

~100 largest CATH
families account for ~30%
of domain sequences

Evolution of protein families.
Annu Rev Biochem, 2005.
Schematic representation of the classification
of sequences from complete genomes into
protein and domain families. The numbers of
domain families identified are given together
with the proportion of domain sequences in
completed genomes that can be assigned
to each type of domain family (CATH, Pfam,
NewFam).

Who is the audience? Expert researchers and those interested in entering
the field.

How will it be used? As a figure in a review article.

What is the goal? To illustrate the steps of an algorithm that categorizes
protein structures, and to summarize the results of the analysis.

What is the challenge? To understand the individual steps.

Suggestions
Use composition to organize information, rather than drawing circles.
Align drawings to the right column edge and give groups ample white space.
Show each step of the cycle. The algorithm uses domain composition to
classify proteins; therefore include domains in the schematics during the
classification step. Assign color to one feature. In the original, color
designates both protein families and individual domains. Use it only for
protein families, and use shape to classify domains.

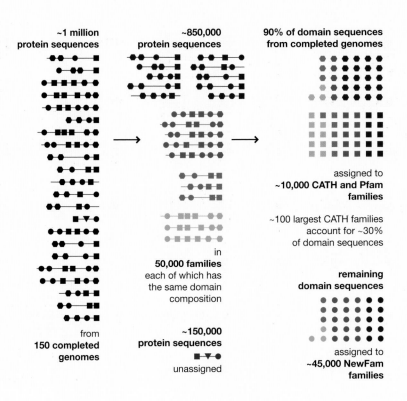

~1 million protein sequences

from **150 completed genomes**

~850,000 protein sequences

in **50,000 families** each of which has the same domain composition

~150,000 protein sequences unassigned

90% of domain sequences from completed genomes

assigned to **~10,000 CATH and Pfam families**

~100 largest CATH families account for ~30% of domain sequences

remaining domain sequences

assigned to **~45,000 NewFam families**

Graphical Tools:

●	COMPOSE	Show steps in columns left to right; depict protein families as clusters in columns.
●	ABSTRACT	Represent all proteins as strings of domains (different shapes).
●	COLOR	Use color to distinguish protein families.
	LAYER	
●	REFINE	Align labels with right column edges; decrease arrow sizes; emphasize relevant information in boldface; eliminate circles around sets.

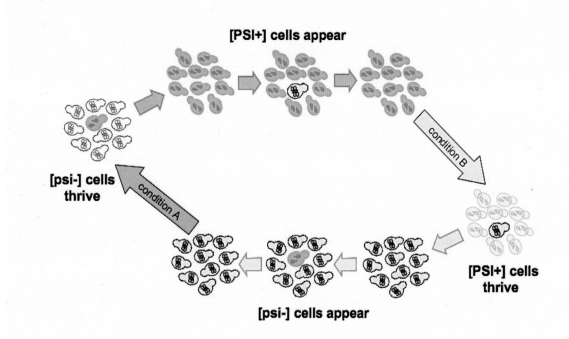

[PSI+] cells appear

condition B

[psi-] cells thrive

condition A

[PSI+] cells thrive

[psi-] cells appear

Yeast prions. *Science,* 2010.
A population of color-coded cells cycling through time and different conditions depicting the appearance and disappearance of protein aggregates in yeast, known as yeast prions. Yeast cells with the prion are [*PSI* +] (red), without it they are [*psi* -] (white). The authors wish to show how [*PSI* +] and [*psi* -] cells can spontaneously appear and disappear, and spread through the population when conditions are favorable.

Who is the audience? Researchers and advanced students.

How will it be used? In a research article and in scientific presentations describing the work.

What is the goal? To understand how cells with and without prions arise, and which conditions favor which type of cells.

What is the challenge? To distinguish the two influences on the type of cells that are present.

Suggestions
Use background colors to organize information. Depict the two conditions affecting the cells as different backgrounds on the left and right, rather than labeling the transition (which was on a diagonal). Abstract each step of the cycle with [*PSI* +] and [*psi* -] cells appearing and disappearing. Use an inset to show the differences in protein aggregation between [*psi* -] and [*PSI* +] cells, rather than depicting them inside each small cartoon of a yeast cell.

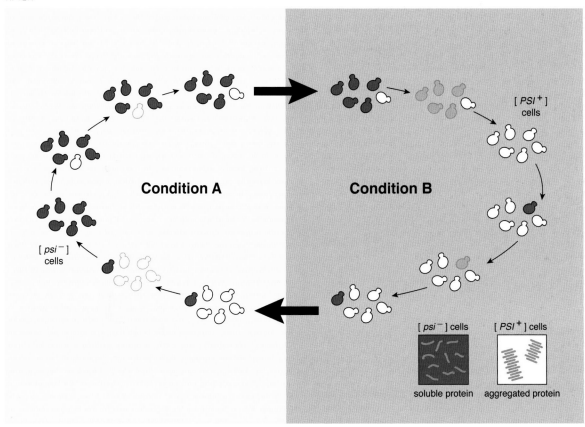

Condition A

Condition B

[*psi*⁻]
cells

[*PSI*⁺]
cells

[*psi*⁻] cells   [*PSI*⁺] cells

soluble protein   aggregated protein

Graphical Tools:

●	COMPOSE	Reorganize the figure to read left to right; put transitions between environments in the middle of the page.
●	ABSTRACT	Show each step of the cycle. Add an inset to show protein aggregation.
●	COLOR	Show the two environments as different background colors.
	LAYER	
●	REFINE	After redrawing, delete unnecessary labels.

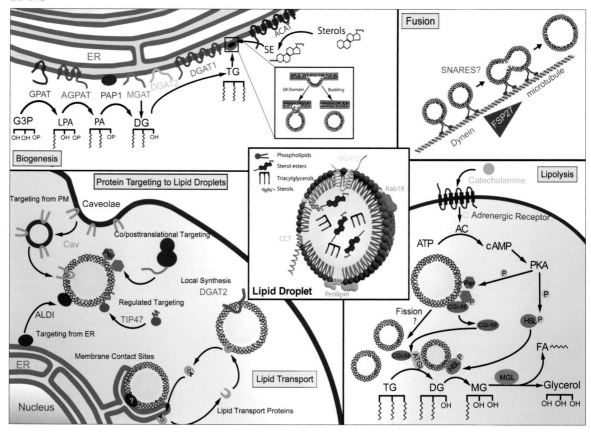

Lipid droplets. *Cell*, 2009.
Lipids are critical for multiple biological processes, and are stored in droplets in the cell. How droplets are formed, how they grow, and how they are used is an active area of research. This figure was submitted for a series called "Snapshots" for *Cell*, summarizing current topics. AFTER is a detail of the final image, which was required to be vertical. Redesign by Andrew Tang.

Who is the audience? Researchers and advanced students.

How will it be used? As a summary figure in a review article.

What is the goal? To summarize information about known molecules and the processes in which they participate.

What is the challenge? To make the graphic comprehensive, but not overwhelming.

Suggestions

Present multiple elements in a unified context. Depict various processes simultaneously inside a single cell. Separate different processes spatially and with different background colors. Use muted and related background colors to group information. In the revised image, pinks and purples are used and are minimally distracting. Use insets where additional detail is required. Here, details of the structure of a lipid droplet, and the processes of docking, budding, and fusion, are shown in insets.

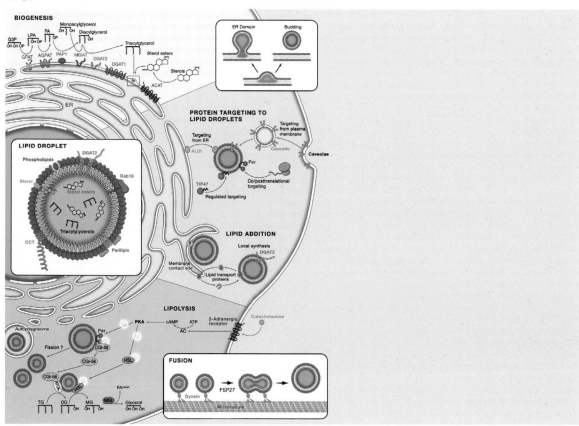

Graphical Tools:

●	**COMPOSE**	Place all processes in a single cell and spatially separate in sections.
●	**ABSTRACT**	Use distinctive cartoons for molecules and organelles in the cell.
●	**COLOR**	Use muted related colors to minimize distraction.
●	**LAYER**	Add insets for necessary details.
●	**REFINE**	Label molecules in corresponding colors and processes in black.

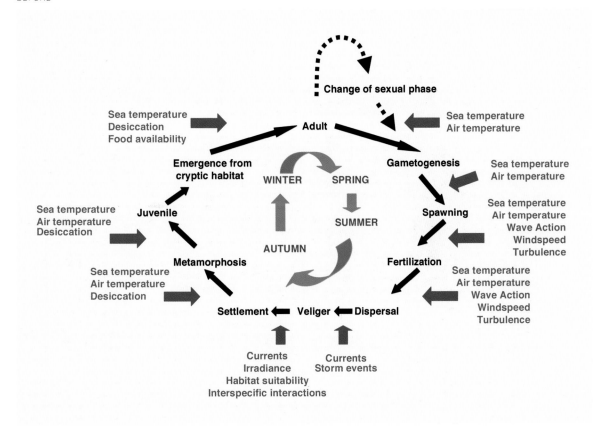

Effects of various factors on the life cyle of _Patella depressa._
Annu Rev Ecol Evol Syst, 2006.
Variable importance of multiple climatic and nonclimatic factors on physiological performance and survival of _Patella depressa_ during different stages of the limpet's life history. Redesign by Glenda Mahoney.

Who is the audience? Researchers and advanced students.

How will it be used? As a summary figure in a review article.

What is the goal? To relate environmental factors and seasonal changes to a complex life cycle.

What is the challenge? To clearly correlate the shared factors across stages and seasons.

Suggestions
Use background colors to organize information. Place the two cycles on a circle where the seasons can be indicated by background colors that emphasize the cyclical nature of the diagram without the need for distracting arrows. Eliminate unnecessary words. Create icons which are more compact and more distinct than text.

PROCESS AND TIME

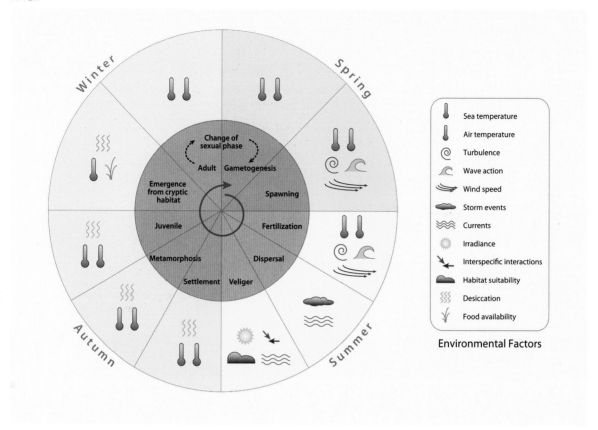

Environmental Factors

Graphical Tools:

●	**COMPOSE**	Organize the life cycle and seasons on a circle.
●	**ABSTRACT**	Show the environmental variables as icons; indicate life stages in text.
●	**COLOR**	Use background color to indicate season; use colored icons to facilitate comparison.
●	**LAYER**	Indicate the life cycle stages in a tinted circle on top of the seasons.
●	**REFINE**	Delete unnecessary arrows.

COMPARE AND CONTRAST

COMPARE AND
CONTRAST

To understand relationships between structures, forms, and processes we are required to make comparisons. Some comparisons are qualitative—for example, the goal may be to understand that two structures are different in size. But how different? A quantitative comparison attempts to answer that question numerically—for example, making it clear that one has precisely half the area of the other. A comparison can also reinforce a characteristic of the primary character in your figure. Simply by including the "other," you can emphasize, by contrast, what a component is or is not.

A fundamental challenge in making visual comparisons is to show the reader that she is seeing a comparison and not just a "list" of things. You must make it clear how to relate the components to one another. This can be reflected in the composition of a figure, where the components are arranged according to some shared characteristic. This strategy is used in the figure at the right, which compares pigmentation patterns on the wings of multiple fruit fly species. Two techniques are used to encourage the reader to consider both the pigmentation pattern and the evolutionary relationship among the species. At the top of the figure, spots unique to one species are indicated by arrows. Because these are shown to be of particular interest, the reader is drawn to look for them in the other images. In the bottom panel, the images are organized according to a phylogenetic tree. This makes the comparison of pigmentation patterns meaningful in terms of the evolution of these traits. In this chapter, you will see how composition and other graphical tools such as color and abstraction can all be used to encourage the viewer to make the intended comparisons.

COMPARE AND CONTRAST

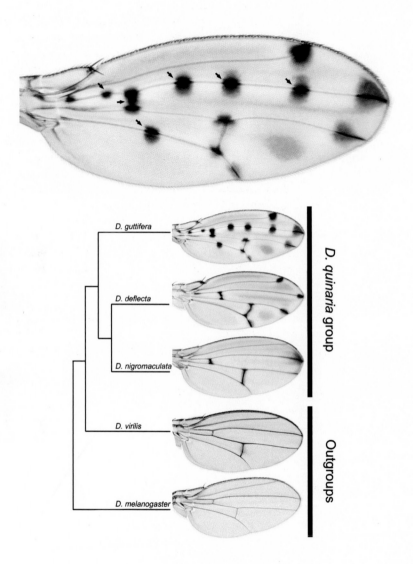

The *D. guttifera* wing exhibits a complex pigmentation pattern. *Nature,* 2010.

Top The wings of *D. guttifera* display sixteen vein-associated spots and four intervein shades; six vein-spots are associated with campaniform sensilla.

Bottom Phylogenetic tree showing the relationships of species bearing different elements of the wing pigment pattern (crossveins, longitudinal wing tips, and campaniform sensilla). The *D. guttifera* pattern is the most complex in that it bears the most pattern elements.

Figure by Thomas Werner.

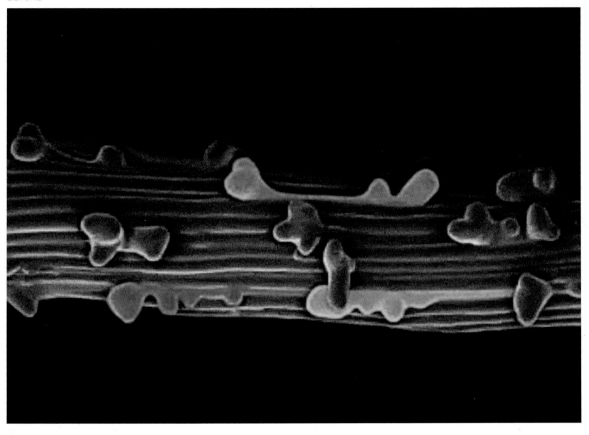

Growth in root hair cells. *Nature*, 2005. Scanning electron micrograph of root surface, showing sites of ectopic growth on the outer face of hair cells. Some hair cells develop multiple sites of growth along their entire length (highlighted in green, orange, and yellow).

Who is the audience? Experts and curious scientists from other fields.

How will it be used? As a figure in a research article.

What is the goal? To point out two different patterns of ectopic growth.

What is the challenge? To show the reader what to compare.

Suggestions

Color according to groups. In the original, the authors used three different colors to indicate three examples of one type of growth, while the examples of the other type are left uncolored. This encourages comparison among the three individuals, rather than between the two groups. We suggest making all examples of one type the same color. This gives both groups the same visual weight, and immediately emphasizes that the groups are the important feature.

 COMPARE AND CONTRAST

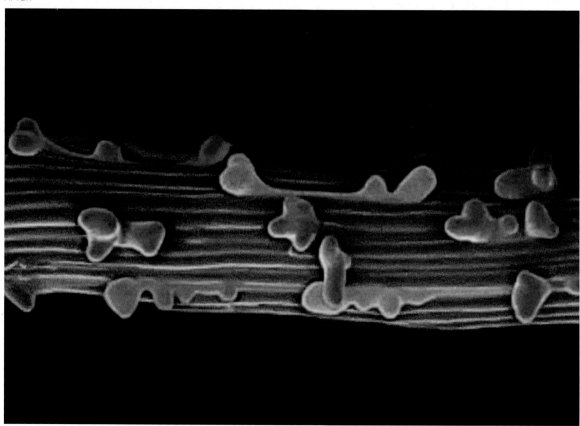

Graphical Tools:

	COMPOSE	
	ABSTRACT	
●	COLOR	Make all members of each group the same color.
	LAYER	
	REFINE	

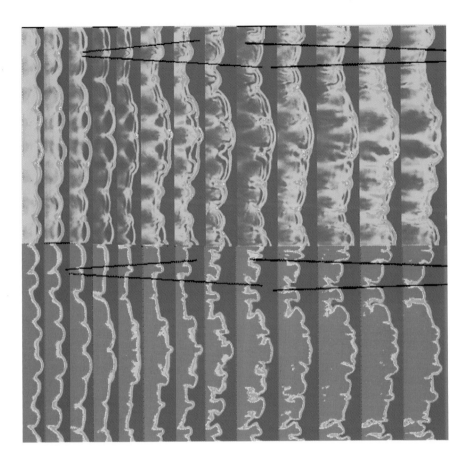

Modeling of accidental explosions.
Annu Rev Fluid Mech, 2004.
Propagation of detonation in two dimensions, showing the developing cellular structure in terms of pressure (top) and reaction progress variable (bottom).

Who is the audience? Researchers and advanced students.

How will it be used? As a figure in a review article.

What is the goal? To see how two variables relate during a process unfolding over time.

What is the challenge? To highlight a particular part of the data, while retaining context.

Suggestions
Clearly indicate the region of interest. Highlight just a portion in color. The reader can then immediately focus on one relevant portion of the figure while retaining an overall sense of context. No information is lost—both grayscale and the heatmap color scheme relay quantitative information. Highlighting in this way makes it easier to compare the top and bottom panels, and makes it obvious that the two shapes the authors indicate are not actually identical.

 COMPARE AND CONTRAST

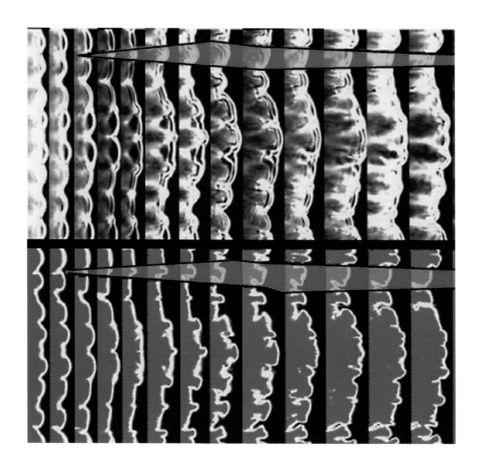

Graphical Tools:

●	**COMPOSE**	Time proceeds left to right; two parameters are shown in series in top and bottom panels.
●	**ABSTRACT**	2D plots over time show simulation values for two parameters.
●	**COLOR**	Use color to indicate values for a subset of the data; use grayscale to indicate values for the remainder of the data to retain context.
	LAYER	
	REFINE	

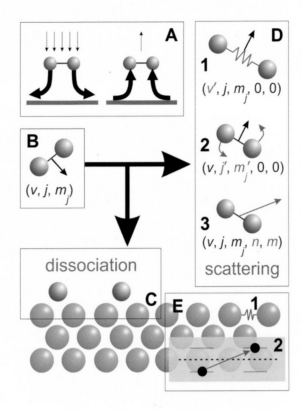

Surface scattering simulations.
Science, 2008.
Schematic representations of some of the processes that can be probed in the scattering of a diatomic molecule from a reactive surface. A molecule approaching the surface in a specific initial internal state characterized by the parameters v, j, and mj is shown.

A The molecule reacts with the surface.

B The molecule scatters back to the gas phase in a vibrationally excited state (1), or a rotationally excited state (2), or with a change in momentum (3).

C All of these processes can occur in concert. Revision by Katharine Sutliff.

Who is the audience? Researchers from a broad range of backgrounds.

How will it be used? As a summary figure for a "News and Views" article, summarizing the results of a research article for a more general audience.

What is the goal? To relate multiple molecular processes that change parameters v, j, and m.

What is the challenge? To focus attention on similarities and differences.

Suggestions
Use an intuitive composition. In the original, the starting point is not clear. The panels are labeled, but they move counterclockwise, then jump to another column. Instead, organize the interactions from left to right. Present all processes in a unified context. In the original, the relationships among the reactions are unclear. Relationships can be clarified by showing all of the interactions on a single surface.

⊖ COMPARE AND CONTRAST

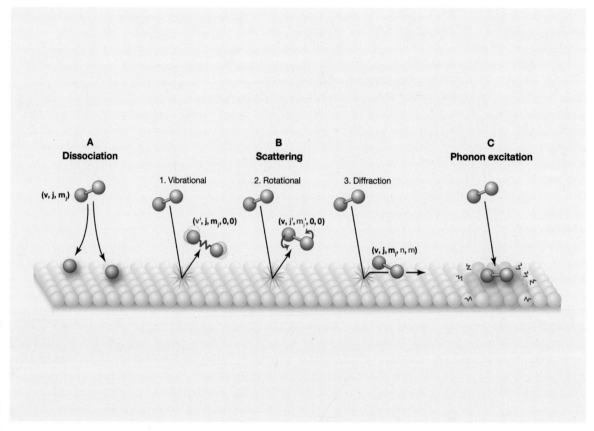

A
Dissociation

B
Scattering

C
Phonon excitation

(v, j, m_j)

1. Vibrational

2. Rotational

3. Diffraction

$(v', j, m_j, 0, 0)$

$(v, j', m_j', 0, 0)$

(v, j, m_j, n, m)

Graphical Tools:

●	COMPOSE	Reorganize to read left to right; show all interactions on the same surface.
●	ABSTRACT	Molecules are represented by balls and sticks.
●	COLOR	Interacting molecules are purple; surface molecules are yellow.
	LAYER	
●	REFINE	Eliminate boxes; group information to eliminate unnecessary panels; add text labels for types of interactions.

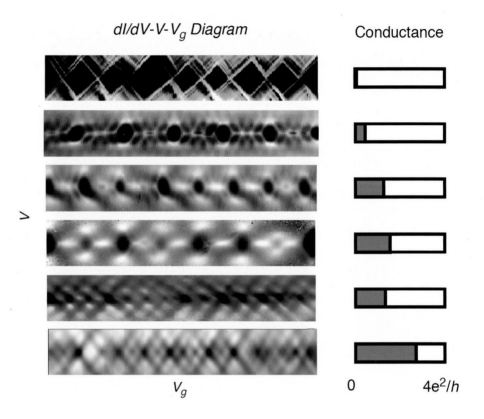

dI/dV-V-V_g Diagram

Conductance

V

V_g

0      $4e^2/h$

**Transport spectroscopy of
chemical nanostructures.**
Annu Rev Phys Chem, 2005.
Six devices are plotted showing various
patterns in a particular measurement of
conductance.

Who is the audience? Expert researchers and those interested in entering
the field.

How will it be used? As a figure in a review article.

What is the goal? To compare the patterns produced by the six devices.

What is the challenge? To clearly show the connection between the patterns and levels of conductance.

Suggestions
Eliminate unnecessary color. Data are already spatially separated in the
composition and do not need to be colored. Simple grayscale images allow
the viewer to quickly see different patterns. Eliminate unnecessary frames
around the bar graph at the right; this facilitates comparison across the rows.

COMPARE AND CONTRAST

dI/dV-V-V$_g$ Diagram   Conductance

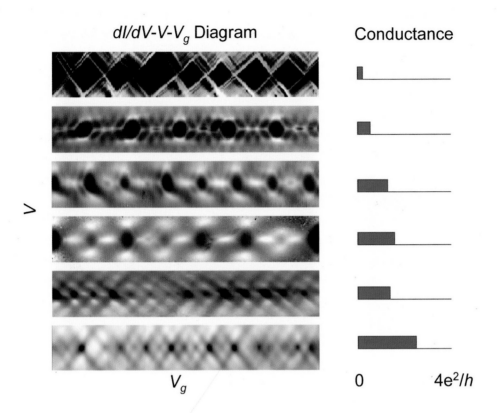

V

V$_g$ 0   4e^2/h

Graphical Tools:

COMPOSE

●	ABSTRACT	Original data are grayscale 2D plots with associated bar graphs to show relative conductance.
●	COLOR	Do not add color to original grayscale data.
	LAYER	
●	REFINE	Remove unnecessary frames from bar graphs.

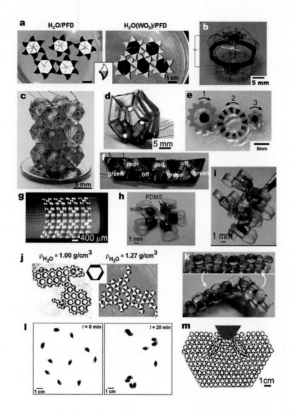

Self-assembled structures.
Grant submission, 2002.
A variety of self-assembled structures shown at different scales. Each panel is labeled with a letter. Original image had a caption explaining each panel.

Who is the audience? Reviewers for a funding agency.

How will it be used? A grant application.

What is the goal? To demonstrate the ability to fabricate a variety of self-assembled structures.

What is the challenge? To meaningfully relate structures within a large range of sizes to one another.

Suggestions
Present multiple elements in a unified context. These devices are dramatically different in size. Organizing them by size helps the reader to relate each to the others. Different criteria can also be chosen (for example, their function or material). Don't discard color if it is informative. Some source photos are in color, which provides additional information about function. Reconsider labeling schemes. Labels can clutter and obscure data. By using a schematic of the drawing, labels can be separated from the photos, but each panel can still be referred to easily.

 COMPARE AND CONTRAST

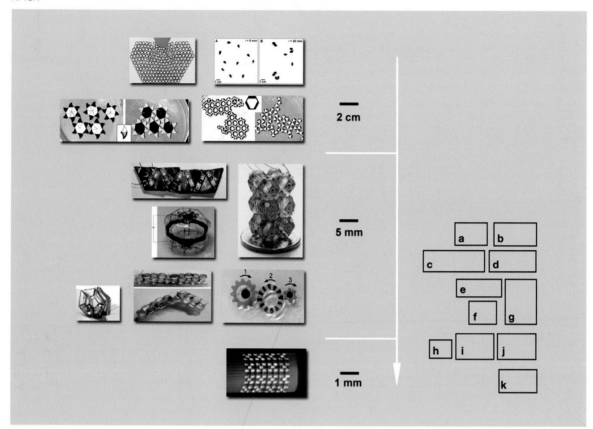

Graphical Tools:

● **COMPOSE** — Organize photographs by size; add a single scale reference.

ABSTRACT

● **COLOR** — Reinstate original color photos when available; use a colored background to highlight data.

LAYER

● **REFINE** — Use a schematic to label components to minimize clutter.

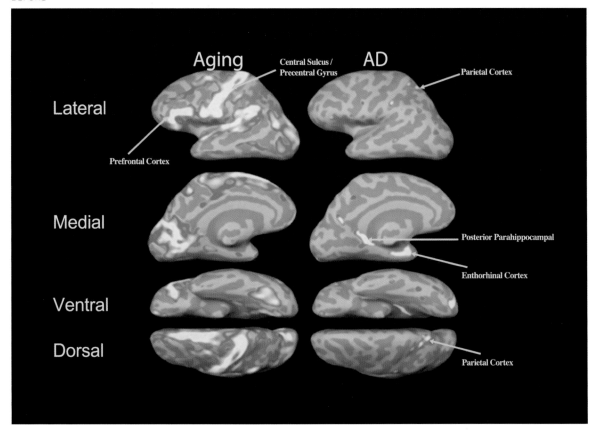

Regional thinning in Alzheimer's disease.
Presentation, 2010.
In Alzheimer's disease (AD), certain areas of the brain lose neurons, or "thin." The authors compare thinning regions in a normal aging patient (left) and a patient with AD (right) by using imaging technology. The study was used to support the need for accurate and automated methods for measuring the thickness of the human brain as a diagnostic tool.

Who is the audience? Physicians, hospital administrators, and funding agencies.

How will it be used? As a slide in a presentation about imaging as a diagnostic tool.

What is the goal? To compare imaging data mapped on two brains.

What is the challenge? To relate the position and intensity of the color in the two images in each row.

Suggestions
Present multiple elements in a unified context. Because the underlying shape and size of the two brains are the same, it is possible to add a third column where both types of data are overlaid on the same brain. This allows direct comparisons, but requires use of two different color schemes to distinguish the data. We also repositioned the labels to make them more regular, and removed the black background.

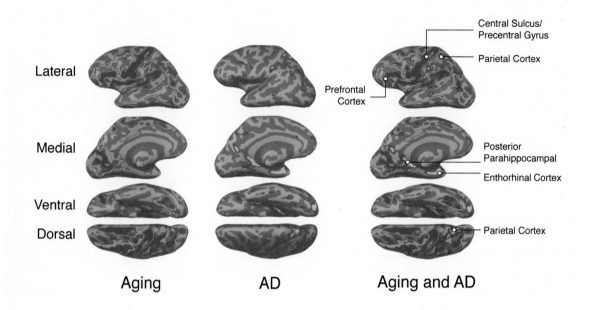

Lateral

Medial

Ventral

Dorsal

Aging   AD   Aging and AD

Central Sulcus/
Precentral Gyrus

Parietal Cortex

Prefrontal
Cortex

Posterior
Parahippocampal

Enthorhinal Cortex

Parietal Cortex

Graphical Tools:

●	COMPOSE	Add a third column on the right to show layered data.
●	ABSTRACT	Data are rendered in color on 3D models of the brain.
●	COLOR	Use different color schemes for data from the normal and AD brains.
●	LAYER	Layer data from both brains on the same model.
●	REFINE	Reposition labels to be less distracting.

CASE
STUDIES

In the world of art, the word "pentimento," which translates literally as "regret," refers to an artist's alteration of a piece during the artistic process, showing traces of previous ideas for a particular work. For example, one can see traces of Da Vinci's charcoal sketches under his painting *Madonna of the Rocks*, indicating that he changed his mind concerning elements of the composition of the piece.

This chapter highlights the process of creating. Various researchers walk us through the process of creating their graphics and talk about the changes they made along the way until they arrived at what they believed was the best iteration.

We believe that "seeing" how others think is a valuable approach for rethinking your own work. Creating a visual representation should be just as much about process as it is about the science. As you follow these short accounts, you might disagree with how the researcher arrived at the various steps, but in their story you might also find a new perspective you may not have considered. And even if the subject matter is different from your scientific interests, we ask that you keep in mind that most of the challenges in representing science are universal and it is more likely than not that you will find a relevant connection.

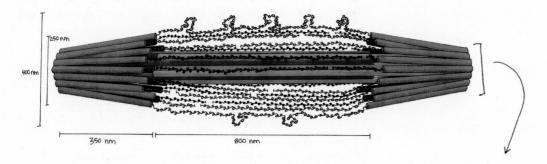

250 nm

400 nm

350 nm

800 nm

Spindle Pole Body
Layout

Entire Spindle

3cm = 100 nm

150 nm

1 cm = 10 nm

⬛ interpolar

CASE
STUDIES

A drawing of the mitotic spindle, by Kendall McKenzie, was part of the collaborative
process described on page 102.

Visually Quantitative Venn Diagrams *by Marc Lenburg*

Marc Lenburg specializes in bioinformatics, a field where Venn diagrams are often used to depict overlap between sets of genes expressed under different conditions. Here he explores how to create a visually quantitative Venn diagram, where the areas of overlap are proportional to the numbers they represent.

Venn diagrams (Venn, J., 1880, see Visual Index) were popularized by Euler in the eighteenth century (Baron, M., 1969, see Visual Index) to graphically represent logical syllogisms, which are categorical arguments such as: no reptiles have fur; all snakes are reptiles; therefore, no snakes have fur. These arguments can be represented by the overlap between categories, but they are not quantitative—note that we don't need to know how many reptiles are snakes in the above example.

However, today Venn diagrams are often used to convey the specific quantity of the unique and shared members among two or more groups. I encounter them frequently in studies of gene regulation, where the overlap between sets of genes expressed under different conditions is depicted, for example, genes that have elevated expression in patients with a particular disease and the genes that are repressed by treatment with a drug. Using Venn diagrams to convey this type of quantitative information can provide a quick summary of the data, but the initial formulation of Venn diagrams (which used circles of equal sizes with a fixed region of overlap to represent the logical relationship between sets) needs to be modified for these diagrams to efficiently convey quantitative information.

A Venn diagram (for example, Figure 1A) displaying the relationship between two groups conveys three quantities: the number of group members that are unique to Group A, the number that are shared between Group A and Group B, and the number that are unique to Group B. The utility of a graphic to convey just three quantities is debatable, but Venn diagrams are popular because they efficiently communicate what the three areas represent in a way that might otherwise be cumbersome to describe in words.

The problem with using the traditional Venn diagram to convey quantitative information is that it does not visually represent the magnitude of these quantities. For example, in Figure 1A, the number of genes detected by "Method A" can be determined by summing the number of group members that are unique to Method A and the number that are shared between Method A and Method B, but this is not communicated visually by this traditional Venn diagram. This is a shortcoming that is quite easy to fix: scale the size of the circles in the Venn diagram so that their area is proportional to the relative size of each of the groups, and position the circles relative to each other such that the area of the overlapping region is proportional to the number of group members that are shared between the two groups (Figure 1B). Doing this just requires a little geometry to figure out the area of the circular segments defined by the chord where the two circles intersect. This simple reformulation of the Venn diagram allows it to communicate both the relationship and magnitude of the quantities being displayed much more efficiently than the standard formulation and more efficiently than could be accomplished in words, and makes the decision to use a graphic to display these quantities more compelling.

It is common to see Venn diagrams that display the relationship between more than two groups, three-group Venn diagrams, for example, are especially popular. Given that it is possible to arrange three circles such that the overlap between all of the pairs is accurate, I wondered whether it was possible to also make the area of the three-way intersection accurate. As a first step, I printed a bunch of three-group Venn diagrams on heavy paper, cut the sections apart, and weighed them on a balance. Surprisingly, most were within a 10% error, which I thought was pretty good given that I was using scissors. However, the more I thought about it, I realized that making the pair-wise overlaps accurate determines the extent of the three-way intersection of the circles without taking into account how much or how little overlap there actually is between the three groups (Figure 2). Despite this limitation with three-group Venn diagrams, the pair-wise strategy of varying the size of the circles and their position can nevertheless provide a visual approximation of the relationship between three groups that can be useful if the actual overlap between the three groups is not extreme. Switching from a generic Venn diagram to the more quantitative version allows the viewer to more quickly intuit the sizes and relationships between groups, as shown in Figure 3.

Figure 1. Gene expression detected by two methods for high-throughput sequencing.

A Traditional Venn diagram displaying the number of genes whose expression is detected by two different sequencing methods. The size of the circles is the same, and their degree of overlap is generic.

B A quantitative Venn diagram that scales the size of the circles to represent the number of genes detected by each method and arranges the circles relative to each other such that the area of overlap represents the number of genes that are detected by both methods. The quantitative Venn diagram highlights that Method B detects the expression of more genes and that most of the genes detected by Method A are also detected by Method B.

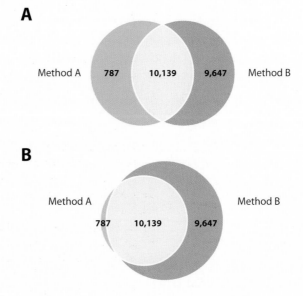

Figure 2. The challenge of quantitative three-group Venn diagrams.

A It is possible to arrange two circles representing Group I and Group J such that the areas of the overlapping and non-overlapping regions represent the number of shared and distinct members between them.

B, C It is possible to add a third circle representing Group K and place it such that it accurately represents the overlap between Group J and Group K, and also accurately represents the overlap between Group I and Group K.

D But placing the circles such that they accurately represent each of these pair-wise relationships fixes the position of all three circles without taking into account the extent to which the three groups overlap (consider that in this example there can be no members that are shared by all three groups).

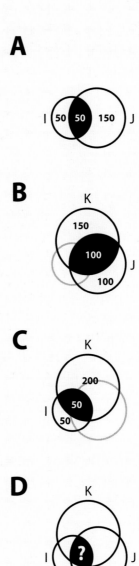

CASE STUDIES

Figure 3. Ability of different microarray platforms to detect changes in gene expression. Overlap of top 2000 genes most differentially expressed between brain tissue and heart tissue as measured by three different types of microarrays.

A The original traditional Venn diagram.

B The redrawn diagram makes it more readily apparent that the three types of arrays (U133, HuEx, and HuGene) give largely overlapping results, but that the HuEx and HuGene arrays are more similar to each other than each is to the U133 array. The greater degree of overlap between the HuEx and HuGene arrays is expected, as they use a similar approach to measure gene expression that is slightly different from the approach used by the U133 array.

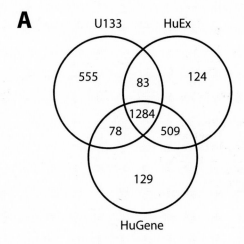

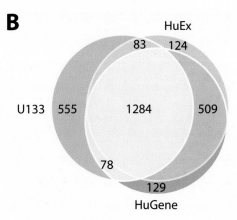

CASE STUDIES

Representing Folded Proteins *by Jane Richardson*

Jane Richardson was asked to review an article for the Proceedings of the National Academy of Science (PNAS). *As part of her critique, she requested revisions to a figure. She submitted her own version of the figure to illustrate her suggestions for improvement. The authors responded with their own version, which she approved and was included in the final article. Although the article was reviewed anonymously, the authors discovered that Dr. Richardson was one of the reviewers (as sometimes happens) and thanked her for her suggestions.*

A This kind of protein structure (or "fold") is called a beta-helix. A beta-helix spirals down like a triangular parking-garage ramp, with parallel beta-sheet hydrogen bonding between the peptides in adjacent rungs of backbone. These helices come in both right-hand (RH) and left-hand (LH) versions; the RH versions (like this one) have one of the three sides bent inward (see the inset cross-section), while the LH versions are symmetrically triangular.

The first task of the figure is to present the overall arrangement of the beta helix. Ribbons are meant to follow the plane of the peptides and H-bonds, so they should lie flat in the plane of the beta-sheet. For some reason, the program used to generate the original figure rendered them as perpendicular to the sheet. As a result, the figure gives absolutely no sense of the triangular-prism shape in 3D. Also, the short helices are over-emphasized, although they are not at all important.

A

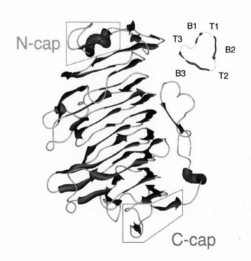

CASE STUDIES

The focus of the paper is to categorize the kinds of "caps" at each end that block this repetitive structure from aggregating end-to-end with other similar molecules. In the original figure, the frames do not define explicitly which residues are assigned as the caps (which are important, however, according to the rules stated in the text), and the colors are not used very consistently or effectively.

B I proposed changes to address these points, while using the same layout and viewpoint. Specifically, the beta-sheets are oriented perpendicular to the axis so that the overall 3D structure (somewhat of a cylinder) is more apparent. And the caps are both shown in warm colors (red and orange), to distinguish them from the main part of the beta-helix while relating them to each other. The labels are in coordinating colors to indicate which parts of the structure they refer to, eliminating the need for the boxes, which were ambiguous.

C The authors responded with their own revision, which was the one chosen for publication. It has rich color and fancy rendering (highlights are excellent, although I feel the shadows are more confusing than helpful), explicit cap regions, and good arrows locally (all in a rotated viewpoint). However, the final figure does not clearly illustrate the overall structure of a beta-helix. In order to understand it, the viewer must already know what a beta-helix looks like in 3D.

B

C

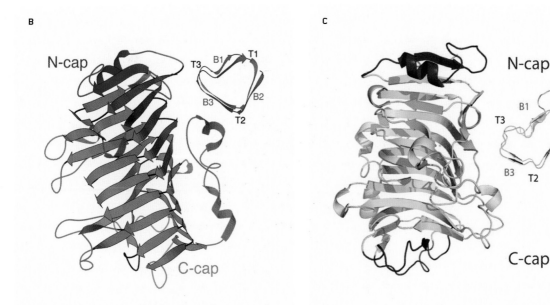

Capturing Time in a Still Image *by Amy Rowat*

Cells grow and divide; they crawl and are squeezed; they alter levels of gene and protein expression in response to environmental cues. While living cells are inherently dynamic, figures are inherently static. This poses a challenge for data visualization. Physicist Amy Rowat presents an example of how to capture time in a picture: a dynamic lineage map that illustrates how protein levels change in a lineage of single cells.

A major goal in biological research is to understand how protein expression in a population of individual cells depends on the age and genealogy of individual cells. We developed a microfluidic device that enables us to track multiple lineages of cells in parallel by trapping single progenitor cells and constraining them to grow in lines within channels for as many as eight divisions. Below is a diagram of our device with its array of channels, designed to simultaneously track multiple lineages. On the right is a detailed picture of one channel: we manipulated the fluid flow to passively trap single cells, so the lineage deriving from a single cell could be contained along the length of the channel.

Having trapped a single cell in a channel, we acquired images at regular time intervals. Over time the cell grew and divided, and filled the channel with cells derived from the single progenitor cell.

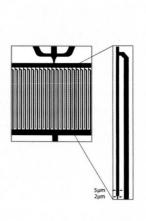

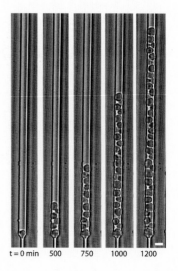

t = 0 min   500   750   1000   1200

CASE STUDIES

In addition, we monitored levels of particular proteins using fluorescent markers and by simultaneously imaging these cells in a fluorescence channel. The challenge was to visualize the protein levels in individual cells in the context of their lineage.

Lineage maps are commonly used to denote familial relationships among a population of cells. By tracking the cells as they grew and divided, we constructed such a lineage map:

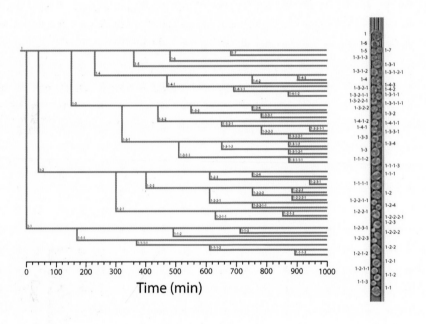

Time (min)

Here the numbers denote the identity of each individual cell in the lineage. But how to represent the fluorescence intensity or protein levels of each individual cell on this map?

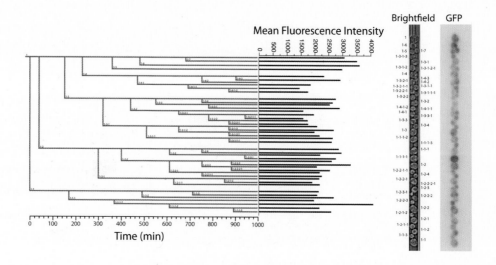

We first tried adding on a bar plot showing the mean fluorescence intensity of individual cells: each blue bar aligns with a red horizontal line representing a particular cell. This successfully showed the variation among protein levels in the lineage; however, this was only a measure of the protein levels at the final timepoint in the experiment. We wanted to be able to visualize how protein levels changed over time as the cells grew and divided.

In order to capture protein level dynamics within the lineage map, we superimposed a color map onto the horizontal line representing each individual cell. Now we were able to visualize the protein levels in every single cell in the context of its lineage. The resultant dynamic lineage map demonstrates the age and genealogy of each single cell, as well as the changes in protein levels over time. Such dynamic lineage mapping can reveal patterns in protein expression that were previously not visible. In this example, cells showed bursts in protein expression, and the bursts in sister cells occurred in a largely synchronized fashion.

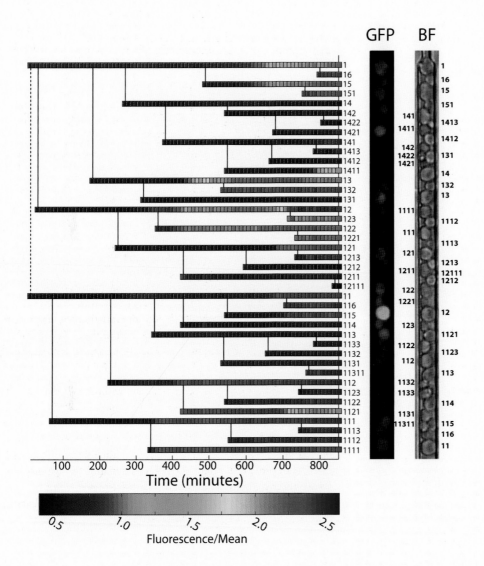

GFP   BF

Time (minutes)

Fluorescence/Mean

Representing Dark Matter *by Richard J. Massey & Lars Lindberg Christensen*

Richard J. Massey (R.J.M.) and Lars Lindberg Christensen (L.L.C.) have a conversation about how they created astronomy's first maps of dark matter. The original unedited dialogue was published in the column "Sightings" from American Scientist, *2007.*

Figure 1

Glowing stars and galaxies make up only one-sixth of the cosmic landscape. The invisible "dark matter" scaffolding is shown here in a two-dimensional map of a patch of the Hubble Space Telescope COSMOS survey. The density of dark matter along each line of sight is determined from gravitational lensing.

R.J.M.   My colleagues and I recently used the Hubble Space Telescope to make the first large-scale maps of the mysterious substance known as "dark matter." Dark matter is completely unlike the familiar material from which everything around us is built—everything that we can see, touch, or breathe. For a start, dark matter is completely invisible! So, aside from the difficulty of finding it in the first place, our map raises a rather tricky data-visualization problem: how to best represent an invisible substance?

L.L.C.   You could compare previous maps of the cosmos—made by mapping luminous galaxies—to nighttime snapshots of a city. Streetlights illuminate some highways and intersections, but most of the interesting neighborhoods remain obscured.

R.J.M.   The universe is exactly the same: the glowing stars, galaxies, and planets make up only one-sixth of the total landscape. The remainder is a grand but invisible skeleton of dark matter. It is the infrastructure of the cosmos, literally holding the universe together.

We can't see dark matter directly, but we can infer its presence via the effect it has on things we can see. So we looked at ordinary but incredibly distant galaxies, which make a wallpaper pattern on the far side of the universe. Light from these galaxies had to pass through any intervening dark matter during its long journey to us. The dark matter shows up in a sort of silhouette against those background lights—but it's not that the distant galaxies look fainter, rather that they change shape. This "gravitational lensing" is like looking through a wobbly sheet of glass. Nearby dark matter gives itself away by distorting the apparent shape of objects behind it.

Figure 2

Three-dimensional map of dark matter, as originally prepared by the scientists. The figure shows an isodensity surface containing volumes of high density; it is oriented with the plane of the sky horizontal and Hubble's view looking down from the top.

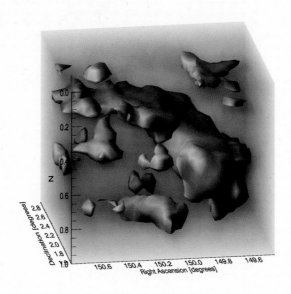

We used Hubble to study the shapes of half a million galaxies, and made a two-dimensional map of dark matter in a patch of sky eight times the size of a full moon. The colors in this were kept dark and shaded blue for a cold, icy feel. Blue is normally avoided in data visualization, because the poor response of the eye makes it difficult to see details, but that seemed wholly appropriate!

The contours were omitted by artist Zolt Levay in a separate version, to provide a smoother, flowing texture (see Figure 1).

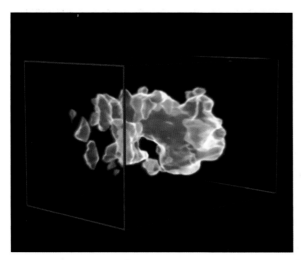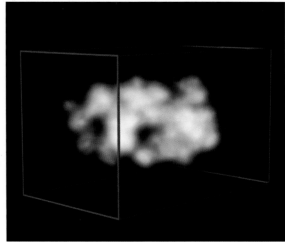

Figure 3
Smoothing and interior light sources were tried to make an isodensity surface for dark matter look more gaseous.

Images on this and opposite page courtesy of Martin Kornmesser, European Space Agency/Hubble.

A tougher visualization challenge was posed by our 3D dark-matter map, which we obtained by also measuring the distance to lumps of dark matter. We made a 32×32×20 data cube, each voxel containing the measured density (a voxel is a cubic unit in a 3D volume, analogous to a pixel in a 2D image). I first drew an isodensity surface, which captured some of the filamentary structure. However, this opaque surface felt too solid and, frustratingly, discarded information about the densest clumps, which were hidden from view. To meet publication deadlines, the figure was scheduled in November for an article in *Nature*. The rest of the COSMOS team ridiculed my apparent pile of deformed potatoes, despairing that such grocery would ever sell to the public. Clearly needing help with visualizations for our January press release, we approached professionals at NASA and the European Space Agency (ESA) (see Figure 2).

L.L.C. Although time was short, and holidays intervened as the American Astronomy Society meeting date approached, we realized the potential of spending time on the data to present this amazing result in a more intuitive manner.

Designer Martin Kornmesser and advanced developer Kasper Nielsen at ESA made another isodensity surface, importing this information as a virtual-reality markup language file into a program used for high-end visualization and movie animation.

R.J.M. We debated a while about geometry. The survey is of a fixed angle on the sky, so it really covers a cone-shaped volume. But showing it as such made it look like the universe is getting smaller over time—whereas the reverse is true. To avoid confusion, we sidestepped the issue and compressed it into a box.

CASE STUDIES

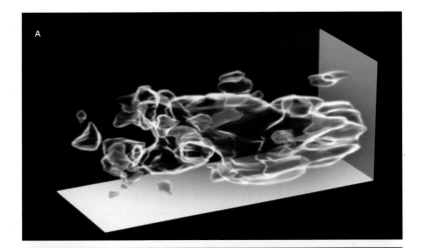

Figure 4

A The designers first rendered the data with glow effects.

B Then a Fresnel shader was used o add transparency.

C Finally context was added: galaxies inside the dark matter, the Universe beyond, and a little Hubble Space Telescope in front. The cool palette intentionally portrays the ethereal nature of dark matter.

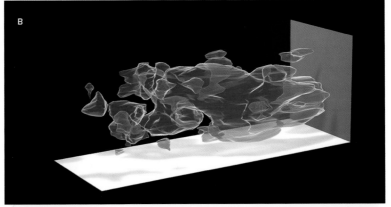

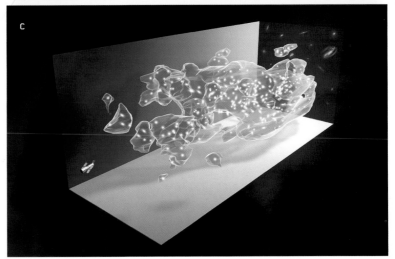

L.L.C. Along the way, we made several attempts to capture the diffuse nature of dark matter, experimenting with various software techniques (see Figures 3 & 4). The rendering was smoothed using a tool for subdividing and refining the representation of surfaces. Light sources were placed inside the isodensity surfaces. We worked with glow effects, and in the end used a Fresnel shader as texture to give the transparent look that helps in getting an overview of the magnificent 3D structure. It was then colored blue. The shader applies the principles of the Fresnel effect, the fact that reflectance depends on the observer's viewing angle. The impression on the eye is similar to looking through a volume of water. We also experimented with context. Clearly the 3D "blobs" look nice but have little connection with the actual Universe. One of our drafts featured the background galaxies, whose photons were so instrumental as "test particles" for probing the gravitational pull exerted by the dark matter.

On top of the dark-matter-density peaks we sprinkled the luminous galaxies, which illustrated another important result: Dark matter and luminous matter are very well aligned, which independently confirms existing cosmological theories. We put in a little model of Hubble itself, out of scale, to indicate how the observations were done. In the end we decided that the way-out-of-scale Hubble and the fake galaxies introduced too much extraneous material, and chucked the draft.

R.J.M. Actually, I really loved that image and have to confess that I slipped it to a couple of journalists. You were worried it looked like frogspawn, but I think that's a lovely analogy for the slippery and elusive dark matter! It shows how well the Fresnel effect works.

L.L.C. Yes, and we'd been working under such intense time pressure that we hadn't realized that you liked it so much. In the end a large package of stills and video footage was released on the web. The main image showed our ice-blue isodensity surface and three planes extracted from the scientists' data with indications of their distances in space. A little Hubble is observing them.

The pioneers in all this were R. Brent Tully at the University of Hawaii and later Margaret Geller and John Huchra of the Harvard-Smithsonian Center for Astrophysics. Tully was one of the first to map luminous matter—galaxies and stars—in the late '80s and early '90s. One of the ways he showed the maps in those early days, when relatively few galaxy positions were known, was to represent their distribution with an isodensity surface.

Of course the resolution of Tully's representations has since improved. This will happen to the dark-matter maps in coming years. Funny how history repeats itself.

R.J.M. There's a natural instinct, when exploring an unknown place, to draw a map. We now know there's a lot of dark matter out there. Science continues the work of the New World explorers, adding flesh to the bones of a map to understand what's around us.

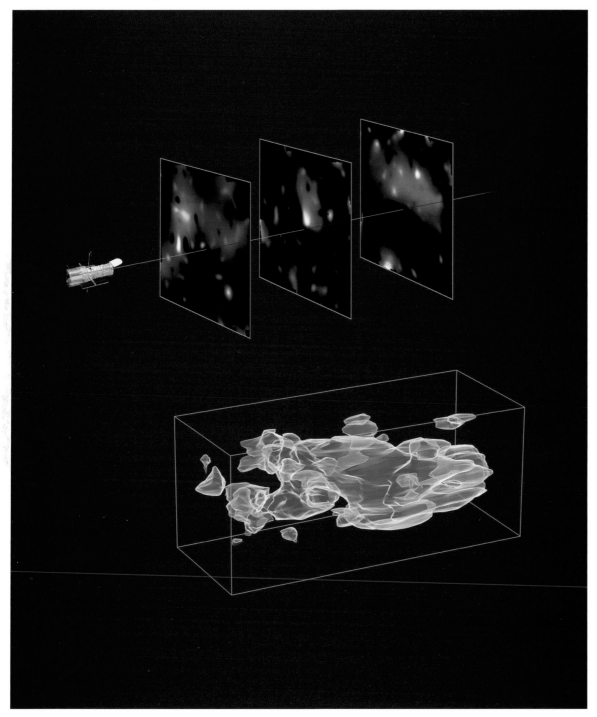

In the final representation of the data released on the web for press and public, three slices above the 3D image show extracts of the data.

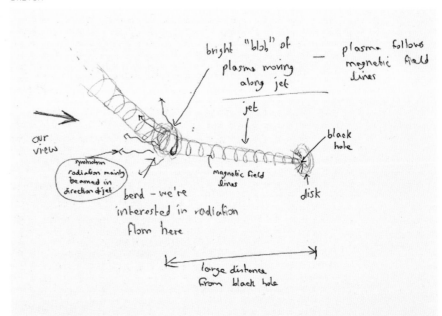

INSPIRATION

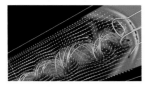

An illustration of a disk and jet from Meier et al., *Science*, 2001. Courtesy of M. Nakamura.

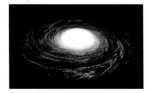

Nature has a library of images, as shown above, for these rushed challenges. Note how it was used in the final graphic.

FINAL DIAGRAM

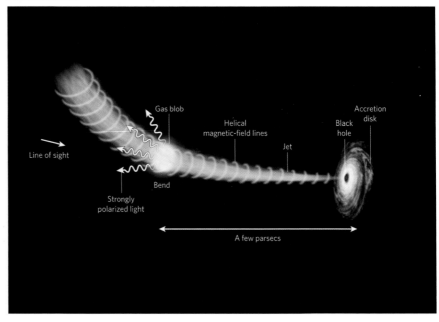

CASE STUDIES

From Sketch to Published Graphic *by Andy Young*

Astrophyscist Andy Young was asked to write a Nature *"News and Views" article about interesting observations of a galaxy that made use of NASA's Fermi gamma-ray observatory.*

This particular galaxy is thought to have a "jet" of material being ejected from its nucleus at close to the speed of light. By looking at how the light from the galaxy varied over time at different wavelengths and how it was polarized the authors worked out where the radiation was coming from and that the jet was likely bent. Along with the article, *Nature* wanted suggestions for an accompanying diagram to explain the research to a non-specialized audience. In this instance, a diagram is particularly useful because the authors arrived at their conclusion by analyzing the properties of the light rather than by directly imaging the system—the galaxy is so distant that its core appears as an unresolved point of light to a telescope. A diagram provides an opportunity to summarize their conclusions in a clear, non-technical way. The turnaround time for "News and Views" articles is short—in this case the article had to be completed in two weeks, and was published in one month. This allowed only a few iterations of the article and diagram.

Before starting to sketch the diagram, I thought about what the paper really wanted to convey to the reader. The key points were 1) the gamma-ray emission is coming from a jet ejected from the nucleus of a galaxy, 2) the jet has a magnetic field, 3) the "flare" they saw came from a bend in the jet that results in strongly polarized light, 4) we're looking down the nozzle of the jet. I produced a sketch of this system (facing page, top left) which had all of these key features, and emailed it to the journal.

My sketch of the disk and jet is a fairly standard way to represent these systems, with the exception of the bend. I also sent some figures that I had found in other journals or magazines: for example, a figure in a *Science* paper showing numerical simulations of jets and their magnetic fields. The art department produced the almost final image in only two days!

The final page proof of the article and diagram was ready the next day, and appeared in *Nature* one week later. I was very impressed by the speed and quality of work produced by the art department on such a short timescale. I think the diagram does a good job of summarizing the important findings of the paper and provides a qualitative understanding of the system as a whole.

Turning Numbers into Graphics *by Alyssa A. Goodman*

Astrophysicist Alyssa Goodman and her collaborators worked to create a graphic demonstrating the success of the "WorldWide Telescope Ambassadors (WWTA) Program" compared to a traditional curriculum. The data included survey results for two groups of roughly 80 students each. Each student, in both groups, was asked the same six questions before and after the "treatment" (either WWTA or traditional teaching), so the data set contains "before" and "after" answers from approximately 160 students. The survey used the so-called Likert scale for responses, with 1 indicating a low value and 5 a high value. In the physical sciences, researchers would typically take these 160 values, combine them into sensible groupings (e.g., by treatment and by before/after), and show as much of the raw data as possible.

When WWTA program leader Pat Udomprasert (a Ph.D. astrophysicist) wanted to explain the data gathered in a survey comparing traditional and WWTA teaching, she created the table shown below (without the color highlighting).

Appendix D: Likert Scale Survey results from Clarke Pilot

Detailed Summary of the pre-test post-test Likert Scale Surveys administered to a group of students who used WWT and a group who did not. Gains that are boldfaced are statistically significant and have a t-test p-value < 0.05.

Likert Scale Questions (1=low; 5=high)	Group A (with WWT)				Group B (without WWT)			
	mean (stdev)		gain	t-test p-value	mean (stdev)		gain	t-test p-value
	before N=75	after N=81			before N=77	after N=75		
What is your level of **interest in Astronomy**?	3.3 (1.0)	4.2 (0.8)	**0.9**	<0.0001	3.7 (1.0)	3.5 (1.0)	-0.2	0.17
What is your level of **interest in Science**?	3.9 (0.8)	4.4 (0.7)	**0.5**	0.0002	3.9 (1.1)	3.8 (1.1)	-0.1	0.45
How much **factual knowledge** do you have about astronomy?	3.2 (1.0)	3.9 (0.7)	**0.7**	<0.0001	3.3 (0.9)	3.6 (0.9)	**0.3**	0.02
How much **understanding** do you have about topics in astronomy?	3.1 (0.9)	3.7 (0.8)	**0.6**	<0.0001	3.3 (1.0)	3.6 (0.9)	**0.3**	0.04
How well can you **visualize** Sun-Earth-Moon relationships?	3.3 (0.9)	4.0 (1.0)	**0.7**	<0.0001	3.7 (1.0)	3.7 (0.9)	0	0.49
How interested are you in using a real **telescope**?	3.5 (1.1)	4.1 (1.0)	**0.6**	0.0006	3.9 (1.1)	3.5 (1.1)	**-0.4**	0.05

CASE STUDIES

Because I am obsessed with visualization, when I saw this (important!!) table, the first thing I did was reformat it a bit, and add green and red highlighting. But then I thought about a tired reviewer's eyes glazing over late at night, and I decided that a summary graphic might be more effective. So, I created the figure shown below, using the "raw" Likert scale values.

It was important to me to highlight the positive/negative differences, which are shown in blue/red highlighting. As a physicist, I am accustomed to presenting as much information about data as possible, so it was also important to me to show the dispersion in the survey results, displayed here as horizontal "error bars" of different line style for "before" and "after." The tiny green type to the right of each bar showed the p-values from t-tests, if anyone was concerned about the statistical validity of the results.

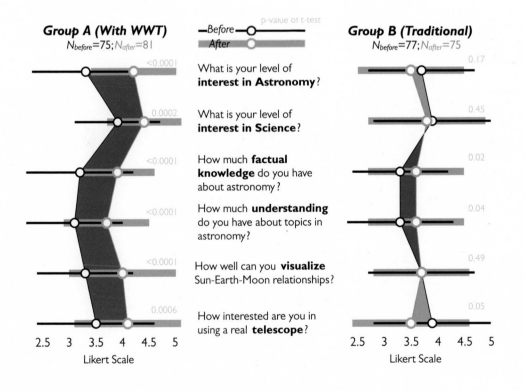

I sent this plot off for comment to the proposal co-investigators, and also to several science education experts who had agreed to advise the group on the proposal/project. The first reaction I got back was from co-investigator Annie Valva, a technology expert at WGBH, the Boston public television station, who wrote:

> Detail: green type barely readable; too light. LOVE format of graphic. Hope it passes muster w/ science types.

But the next reaction was from Philip Sadler, the world-renowned science educator. Phil wrote:

> A visually interesting graph, but I think it may be a bit misleading. You connect each question as one would for a continuous function, but these are discrete questions and your graphic would look quite different if you simply change the order of the questions. Plus, error bars generally show $+/- 1$ standard error of the mean $(SD/n^{0.5})$ when looking at such differences, not the standard deviation of the distribution. You want only to know if the difference seen is statistically significant at some accepted value (0.05 is fine in this case). I include a more boring looking rendition of the data. If you want to employ the SD (standard deviation) in showing your results, for educational research the standard is to show the change in units of standard deviation, the so-called "effect size." I have included a second graph showing this. With effect size, one can use $+/- 2SE$ to see whether the difference is significantly different from 0.0. Effect size is the common metric by which educational impact is compared between studies, 0.25 being a small effect, 0.50 medium, and 0.75 and above is large. So the difference you see is between medium and large for WWT, but your controls show little impact. I enclose my Excel spreadsheets showing both kinds of graphs.

> Answers to your questions: 1. Error bars should not show standard deviations but standard error of the mean and their description should be on the graph. 2. A graphic of gain/loss is fine; one does not need the numbers on the graphic. 3. Significance is generally reported as 3/40.05, 3/40.01 or 3/40.001. This is far less important than the magnitude of the gain/loss and once a given threshold is reached (say, 0.05 for small studies like this one), researchers make little of its value past the threshold.

Phil was correct to suggest that the trend indicated by the undulating blue/red swath could be misinterpreted. He also was generous enough to remake the chart (shown on the adjacent page), in what he called a "more boring looking rendition," to put it more into a context that education (and human-subject) researchers would appreciate and be familiar with. Importantly, Phil introduced the notion of "effect size," which shows a change in units of standard deviation. Using effect size compresses information into a readily interpretable single number, but it also loses information — about, for example, the numerical value of the standard deviation. Sadler also educated me about the standard interpretation of effect size in his letter. Terms such as "small," "medium," and "large" are not often used among physicists, in spite of their practical value! Nonetheless, I was happy to revise the graph to use terminology and formalism that are standard in education.

The final version of the graphic in the NSF proposal describing these survey results is shown on the next page. The information is essentially the same as that presented to me in Sadler's graph, but I made significant graphical changes. The Excel chart created by Sadler repeated columns for "Without WWT" and "With WWT" in such a way that it was not obvious which values were being compared. So the first change I made was to re-orient the chart with the questions listed horizontally, allowing for the "answers" to each question to be sensibly displayed next to each other, along a single horizontal line. I added light-colored connecting lines between answers to

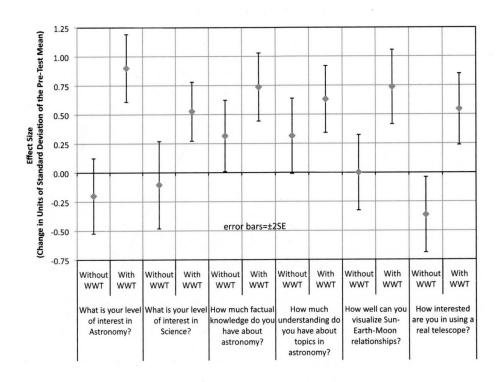

preserve the kind of "difference" effect that had formerly been established (in my first version) by the blue and red shading. The lines I added also solved the problem in Sadler's graphic where it was unclear what was being compared; the lines clearly show which values go with which. I also added a single "zero line" to show where zero effect lies, and I removed all axes except for a single "Effect Size" axis shown at the bottom of the plot.

In addition to the re-orientation and axis modifications, I also made new stylistic choices: the "traditional" and "With WWT" groups were given different colors, and important text was boldfaced. (Note that blue is the color of the WWT logo, and black is boring, so the choices were intended to have a subliminal effect.) I also chose to include the sample sizes in small text below the group labels, and I added an explanation—at Sadler's suggestion—of effect size directly within the x-axis label of the plot.

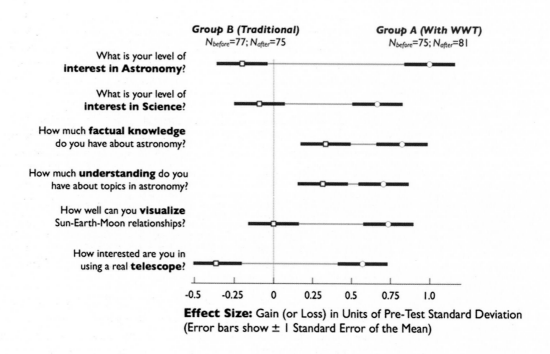

Sadler's reaction to the new plot was:

> Looks good. It conveys the information, but is far more attractive than the conventional graph. I would change the axis label to be Effect Size: Gain in Units of Pre-test Standard Deviation. I would be sure to mention that error bars are ±/1 Standard Error of the mean.

So, in the end, a compromise new plot was achieved. But because this proposal was intended both for education and for physical science researchers, additional explanation to accompany this new plot was needed. Here is the text that accompanies the plot in the proposal:

> As with the Moon Quiz, we surveyed two groups—one that used WWT (Group A) and one that did not (Group B)—about their interest level and self-perceived understanding of astronomy and science before and after the Astronomy unit. We used a Likert scale (1=low; 5=high) on the survey, and we present the survey analysis results in the graph above [here shown on the opposite page] in terms of the effect size measure often used in survey research. Effect size measures the gain (or loss) in units of the pre-test standard deviation. As this is not a familiar statistic in the hard sciences, we remind readers that, conservatively, effect size absolute values of 0.25 or less indicate essentially no effect, 0.25 to 0.5 a small effect, 0.5 to 0.75 medium, and 0.75 or greater large. Each point plotted shows mean effect size, and the bars show ± one standard error on the mean. Raw survey statistics are included in this proposal as Appendix D.
>
> Group A (With WWT) showed statistically significant gains on all questions asked. Group B (No WWT) showed statistically significant gains in their self-reported factual knowledge and understanding of astronomy topics, but they did not show gains in interest in astronomy or science in general. Group A self-reported a significant gain in the ability to visualize Sun-Earth-Moon relationships, while Group B did not, consistent with the results of the Moon Quiz described in Section 3.3.1. One concern expressed about WWT is that the beautiful immersive environment might lead users to lose interest in using real telescopes: our data indicate that the contrary is true (note the last question in the figure).

An interesting postscript: As lovely as the final plot is, whenever the program leader or I show it to our astrophysics colleagues, we have to explain the meaning of "Effect Size." In fact, we often wind up showing the original table to some audiences—and that table is included in the proposal as an appendix, for all those physicists who must have the raw data!

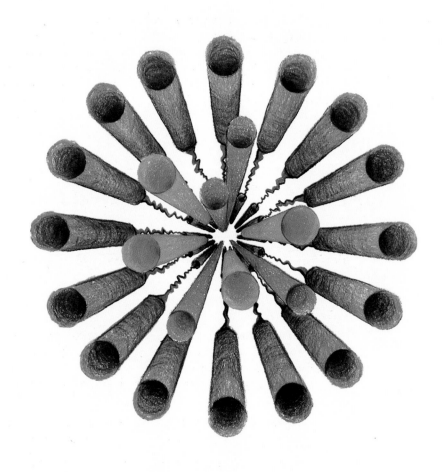

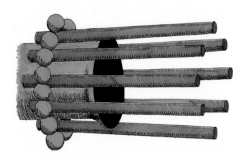

CASE STUDIES

Intersecting Art and Computer Graphics *by Russell M. Taylor II & Kerry Bloom*

A multidisciplinary team of scientists and artists including the authors, Kendall McKenzie, Leandra Vicci, Andrew Stephens, Jolien Verdaasdonk, Steven Nedrud, Mike Falvo, and Shi Fu, worked together for two years to develop a 3D model of the mitotic spindle structure of yeast cells during metaphase (the time in cell division when all of the chromosomes are brought together to line up). The following chronicles the development of the models from scientist-drawn white-board images through scale-model artistic drawings to 3D computer-rendered models.

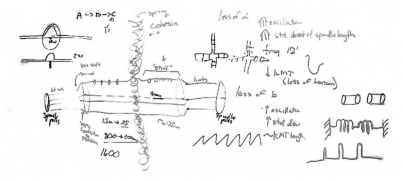

The image above is typical of the drawings we used to communicate with each other for the first eight months. We covered two large white boards with such drawings during each meeting and took photographs to record our progress. Although it was easy to draw them rapidly, these images were deficient in several ways.

- They were often not drawn to scale (making it difficult to estimate chromosome length and preventing correct reasoning about spatial layout).

- They often lacked important details (making it difficult to remember context).

- They were in an abstract projection space (making it impossible to reason accurately about free space and collisions between parts of the model).

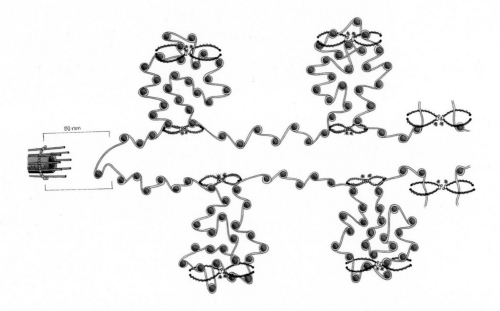

In the ninth month of the collaboration, we were joined by artist Kendall McKenzie, who created hand-drawn images to scale. Her image above shows a detailed view of yellow DNA wrapped around orange histones, tied together by blue condensin and purple cohesin molecules.

Constructing these drawings led to the correction of scientific misconceptions about spindle geometry. The drawing on the top of the following page shows the entire mitotic spindle, with chromosomes attached to green kinetochore microtubules surrounding red interpolar microtubules. When we originally specified the dimensions and angles for this drawing, Kendall pointed out that four of the quantities specified (width at the end, width near the center, length, and angle) were inconsistent. This required a reworking of our model for the geometry of the spindle so that a consistent drawing could be made.

Kendall also constructed a 3D physical model out of pipe cleaners and colored cotton balls that sat in the center of the table and provided a rapidly changeable model to help us think about the spindle's dynamic behavior (photo on the bottom of the following page).

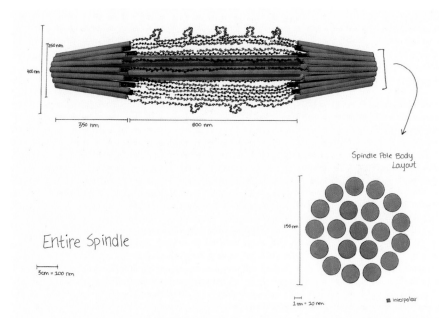

250 nm

400 nm

350 nm

800 nm

Entire Spindle

3cm = 100 nm

Spindle Pole Body
Layout

150 nm

1 cm = 10 nm   ■ interpolar

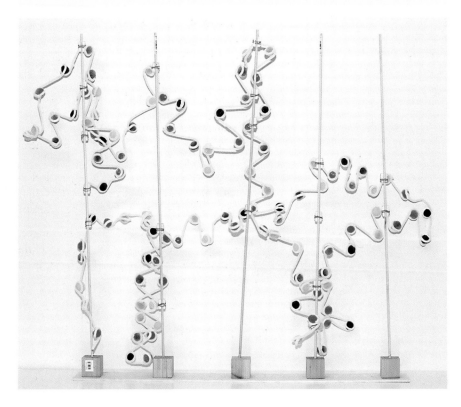

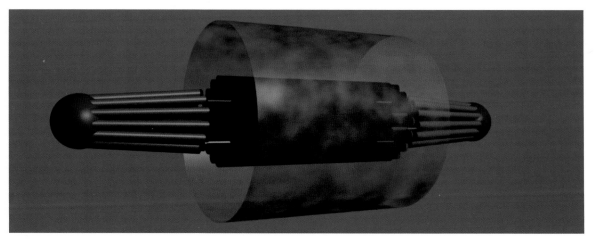

So that we could get an accurate 3D rendering of the structure, and so that we could view it in stereo 3D, we created two 3D computer-generated models of the spindle, first a hand-edited model (in the ninth month) and then a parameterized program-generated model (in the twenty-second month).

The model pictured above was made to test the hypothesis that the chromosomes were organized into a 30nm-wide coiled form, rather than random loops. The distance between the dark central cylinder of chromosomes and the gray experimentally determined cohesin ring showed clearly that this was too compact.

A parameterized model that could be regenerated and rendered automatically gave us a more rapid iteration of the model (the hand-drawn models were regenerated from scratch each week). In addition, we were more able to change parameters and quickly see the results.

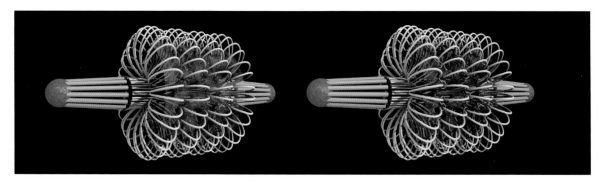

CASE STUDIES

The current working version of the model is shown on the bottom of the adjacent page. Its chromosomes extend beyond the cohesion barrel, it has long strands of unwrapped chromosome near the microtubules, and it has no cohesion on the outer side of the chromosome rings. If you look carefully, you will see purple condensing "staples" near the center; however, the depiction is somewhat misleading because they should not be regularly spaced. This is a typical example of having to make choices for clarity. Otherwise, the representation matches our current understanding of the spindle structure.

The simplified image above was made for inclusion in a paper describing the spindle structure. The removal of all but one chromosome simplified the presentation but didn't show the space-filling packing of the chromosomes. The image was further simplified for publication (not shown) because the style of the rendered drawing was unfamiliar to reviewers, who said that they were confused by the complicated shading nature of the figure. The creation of these models was profoundly useful as a tool to drive our thinking and understanding of the subject.

INTERACTIVE GRAPHICS

Interactive graphics are becoming more prevalent in research and education. Scientists and faculty are using interactive graphics in presentations, coursework, and electronic textbooks. Just as important, many journals are expanding their efforts to create interactive opportunities for their readers. Multiple journals are developing design teams to find new approaches to communicate the enormous quantity of new data researchers are collecting; many are choosing to explore the possibilities offered by interactive graphics.

The online journal *PLoS* (*Public Library of Science*) engages readers with the graphics in their articles in two unusual ways: readers can pick and choose which figures they want to see enlarged directly from the article, and some figures themselves are interactive, wherein readers can rotate and re-orient with their cursors to reveal 3D structures. These particular technologies for *PLoS* were co-developed with MolSoft LLC, iSee, and the Structural Genomics Consortium at the University of Oxford. Brian Marsden at iSEE writes:

> Clearly, the use of deeply embedded and interactive multimedia is not just limited to structural biology. Any three-dimensional data in which predefined visualizations or viewpoints can be captured can be presented in this way. While the technology is doubtless ready, there is a restriction to wider-spread adoption by publishers—publishing systems are almost universally unable to currently support this approach. The future of this form of dissemination of information depends on industry-wide adoption of this new generic technology.

In this chapter, we present several examples of interactive graphics, including animations, from different contexts along with in-depth discussions with their creators, who address the challenges of creating interactive graphics and the tools to tackle those challenges. The principles that guide you to prepare a good static graphic apply equally to creating an effective interactive graphic. Fundamentally, you must still tailor content to your audience and format, and decide what you want your viewer to see first. It is impossible to do justice to an interactive graphic in a static format like this book. Each example in this chapter can be found on our *Visual Strategies* website, which we encourage you visit (see Appendix).

PLoS Biology : Publishing science, accelerating research

http://www.plosbiology.org/article/slideshow.action?uri=info:doi/10.1371/journal.pbio.1000439&imageURI

Return to article

◀ PREVIOUS | NEXT ▶

View Larger Image   Download original TIFF   Download PowerPoint Friendly Image

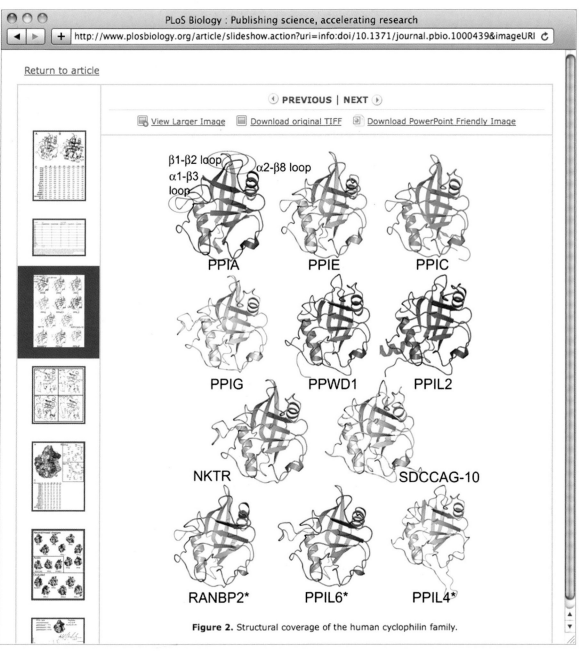

Figure 2. Structural coverage of the human cyclophilin family.

A web page from a *PLoS* article. Readers are able to select a figure or table (blocked in blue) as they read the article to then see it enlarged in a new window. All figures in the article are accessible from this window, making them easy to browse.

Interactivity Brings Clarity *by Kathy Stern and Daniel Muller*

The initial graphic below accompanied a review article on genomewide association studies. When the author submitted his manuscript, the editor at the New England Journal of Medicine *suggested that the journal create an interactive graphic that would be more legible. The author was amenable, so together with Daniel Müller, one of the staff medical illustrators at the journal, they decided what would be most effective.*

BEFORE

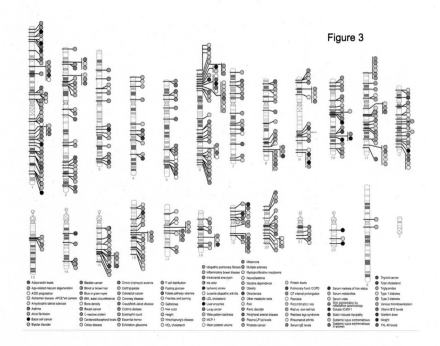

Figure 3

The original static graphic shows the results of all genomewide associations from 545 studies reported up until March 2010. The goal was for readers to be able to see which genomic locations are linked to which diseases—but the graphic is really too complicated to see that easily. The only thing you can readily pick out are the locations where an unusually high number of diseases are mapped. Many graphics that appear in papers are too complex to read easily—especially with all of the high-throughput data that are being generated these days.

INTERACTIVE GRAPHICS

Interactivity can go a long way toward communicating data more clearly. The team first decided to group all the diseases and traits into 7 major categories in order to decrease the amount of data the user is initially presented. They also decided to use the same basic layout as the static graphic, but allow the user to toggle each disease category on and off. This allows the user to filter and focus in only one category or compare multiple categories at any given moment. In addition, the specific name of the disease or trait is displayed as a pop-up label when the user mouses over each individual region. The interactive graphic still uses color for the different categories as well as the

AFTER: ALL ON

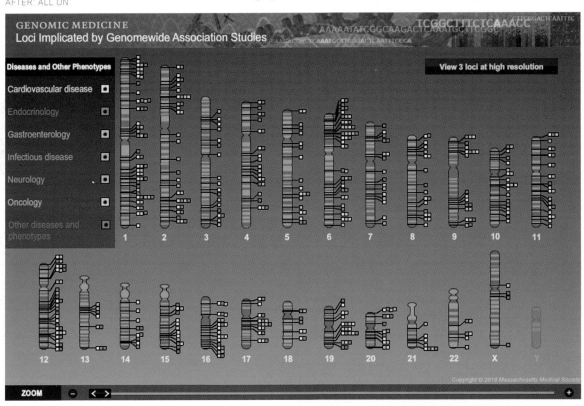

DETAIL

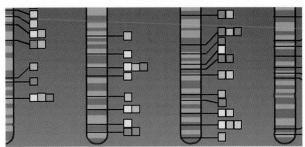

AFTER: NONE SELECTED

AFTER: GREEN & YELLOW SELECTED

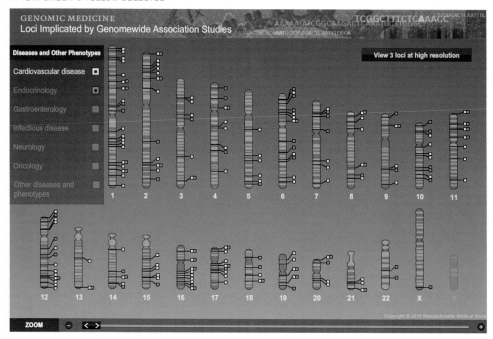

INTERACTIVE GRAPHICS

convention for the human karyotype, i.e., banded chromosomes. A zoom tool on the bottom of the interactive graphic makes it possible to magnify any given part of the graphic and see in more detail chromosome areas where multiple diseases and traits are located very close together. In addition, for just a few representative loci, the user can view detailed information by clicking on a button.

Compared to the original static graphic, an interactive graphic makes it much easier to see how many genomic regions are associated with each category, where they map, and whether they co-occur with regions associated with other categories. However, both the static and interactive graphics are of limited resolution and not linked to the associated primary data. A big challenge is deciding where to draw the line on how much information to include. There is considerable detail available for each of these studies, so the designers could have looked for ways to link additional information. For instance, you cannot determine the size of any of these regions (they can vary quite a bit), or get any further granularity within the categories (for instance, ask how many regions are associated with high LDL cholesterol specifically). But the same basic design framework could be used to address these inquiries; the challenge would be to determine how the data driving the graphic should be managed.

The potential for some of these graphics to have a longer lifetime beyond the paper in which they are published is exciting. If the format is useful enough, the graphic could be associated with a database and new data could be incrementally added, keeping the figure current. This could create a new role for review papers, where some of the infrastructure could be developed and actually made more permanent. However, how to fund these sorts of resources is unclear—neither research funders nor publishers have grappled with this issue yet.

MizBee *by Miriah Meyer*

MizBee is an interactive tool for comparing genomes. It was built as a collaboration between Miriah Meyer, a data visualization researcher at Harvard, and Lijun Ma, a genomics researcher at the Broad Institute of Harvard and MIT. The primary goal was to facilitate Dr. Ma's research on genome duplication. But MizBee has proven useful for other types of comparative genomics research as well. Here, Dr. Meyer discusses the thinking behind the tool.

This project started with a static graphic built to accompany an article that proposed a whole-genome duplication event for a species of fungus. The figure in the article shows regions on each chromosome that are similar to regions on another chromosome, encoded by color (opposite page). Two similar regions are expanded at two different scales in the inset. Note that for this region you see a reciprocal relationship between the olive and purple chromosomes; there is a region of the purple chromosome with similarity to the olive one, and a region of the olive chromosome with similarity to the purple one. This indicates a shared evolutionary history between these regions. In the inset, transparency is used to indicate a larger, inferred region of similarity at higher resolution. Darker blocks are chromosomal regions of the highest similarity; gray lines indicate which blocks are related. Individual genes are not visible in the high-level chromosomal view. But in the inset, they are indicated by arrows below the colored bars. This kind of figure is fairly common in genomics, a field in which researchers try to understand how genes are related to one another, both within and between species.

But the static figure has limitations. The viewer can see only one region at a time, and it is not possible see the relationships of the entire dataset at multiple scales. Lijun Ma was interested in developing a tool to help her communicate her findings for additional genes, and allow her to explore her data more generally. I worked with Lijun to develop an interactive tool using many of the visual conventions from her static representation.

The project's main challenge was how to present the pair-wise relationships of similar regions between different chromosomes. In the literature, two approaches have been commonly used; paired chromosomal regions are indicated with color or with connecting lines. Each approach has its limits. If we choose color, we can't comprehensively label all regions because we can easily distinguish only eight to ten colors at a time. If we choose connecting lines, the graphic becomes quickly overwhelmed by visual clutter when looking across the entire genome.

INTERACTIVE GRAPHICS

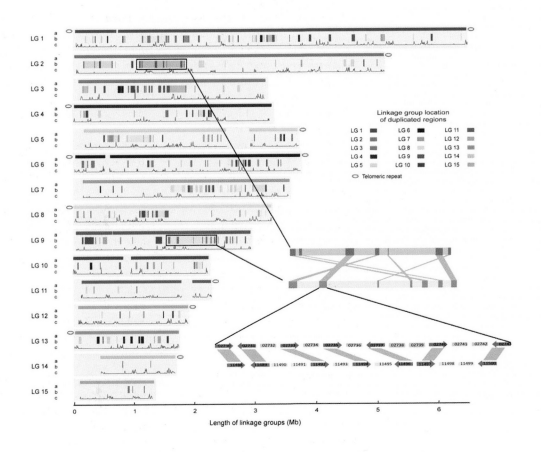

Length of linkage groups (Mb)

INTERACTIVE
GRAPHICS

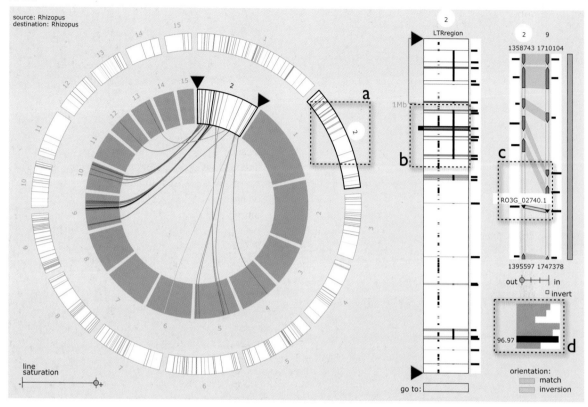

To solve this challenge, we used a redundant encoding of pair-wise relationships (see above). We used lines drawn between regions to show pairs and a repeating eight-element color map to help distinguish regions from one another. We used several techniques to reduce visual clutter. One technique was to use a circular layout for the two genomes. One genome is represented on the internal circle; the other is represented on the outside. Individual chromosomes are shown as numbered blocks with light gray background in between them. The circular arrangement is compact, and reduces edge crossing for the lines, compared to the linear arrangement in the original graphic. This approach allowed us to easily represent similarities within a single genome (colored lines that connect within a circle), as well as similarities between genomes (color blocks between circles). For this particular figure, the researcher is comparing chromosomes within a single genome, so the two circles represent the same genome. Another technique that we used was edge bundling to pull together the lines drawn to similar locations so that we didn't have too many lines on the screen. Finally, we added interactivity that allows users to show connections for only one selected chromosome or region at a time so that they can focus their attention and eliminate unnecessary information from their field of vision.

INTERACTIVE GRAPHICS

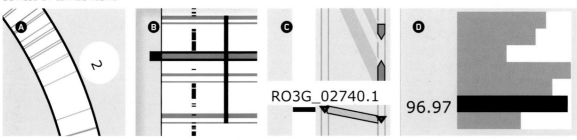

To solve the problem of understanding the data at multiple scales, we employed a common strategy called "linked views" in which multiple different types of representations simultaneously update, based on choices that users make. Each view is optimized for viewing different types of relevant information; in this case the views can communicate detailed information on the location, size, orientation, and degree of similarity of each region. Working from left to right, each view shows more information at progressively higher resolutions of the data. A chromosome is selected by clicking on its numbered label in the outer ring, and both the circular view and the middle view are updated to show that selected chromosome's information (panel A). In the middle view, individual blocks are selected with mouse clicks (panel B), which updates the detailed block view on the right. Rolling over individual genes in the block view creates a pop-up with the gene's name (panel C) and also highlights the corresponding similarity value in the linked histogram (panel D).

One of the most successful things about this project is the fact that researchers were able to explore much more information. Not only could they glean more information from this particular dataset, but also the tool has proven useful for depicting different datasets with similar types of underlying data. Interactive tools like this can replace confusing high-density data presentation in static figures, offering easy and intuitive exploratory interfaces. This particular tool has been used in subsequent work and as supplemental material for reviewers. However, as with any interactive tool, getting it to generate the right static graphic for a presentation or a publication requires additional finessing. It is not as simple as generating a screenshot while using the tool. However, many journals allow submission of the interactive tool itself as supplemental information—bypassing the need for a static representation altogether. A custom interactive tool like this, made for exploring data unique to a researcher's particular work, may be difficult for researchers to implement on their own. Doing so often requires a level of expertise in programming, design, and computer science—a researcher may therefore find it useful to cultivate creative collaborations with computer scientists and visualization researchers, who are always on the lookout for real-world research problems to tackle.

INTERACTIVE
GRAPHICS

Exploring Brain Connectivity with Two-Dimensional Neural Maps
by Çağatay Demiralp

The goal of this project was to provide a representation and mode of interaction that would make it easier for neuroscientists to explore the intricate neural connectivity of the brain. The source of the brain connectivity information used was diffusion-weighted magnetic resonance imaging (DWI).

Figure 1 An example of a tractogram, where tracts are rendered as tubes.

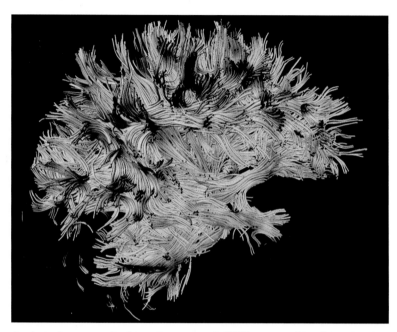

A relatively new medical imaging protocol, DWI, now enables the estimation of neural pathways in the brain as a collection of space curves called a tractogram (see Figure 1). With some stretch of the terminology, each curve in a tractogram is also called a tract. Because DWI is the only widely available non-invasive method providing a window into the connectivity of a living brain, study of tractograms (i.e., tractography) has important applications in both clinical and basic neuroscience research on the brain. These datasets, however, have a visual complexity proportional to the intricacy of the axonal brain connectivity. And, with increasing DWI resolutions, this visual complexity is becoming greater and greater. It is thus often difficult for practitioners to see tract projections clearly or identify anatomical and functional structures easily in these dense 3D curve collections.

INTERACTIVE GRAPHICS

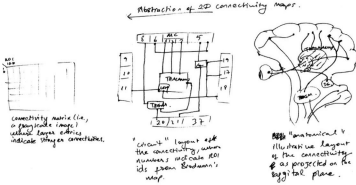

Figure 2 A sketch from our notebook considering different levels of abstractions for 2D brain connectivity representations. Left: connectivity matrix. Middle: circuit layout. Right: anatomical illustration. The degree of abstraction decreases from right to left.

And so, we began thinking about new representations to improve understanding and the exploration of these datasets (Figure 2 shows a snapshot of a thought experiment about this challenge in our notebook). We envisioned that these new, and possibly simpler, representations would be typically used in tools where both the new representation and the conventional 3D tubular tractogram model would be displayed together and user interaction on their views would be linked.

Our first attempt was to use a 2D point-set representation of tractograms, where each tract was represented with a single point in the plane. Placement of the points in the plane reflected the configuration of the tracts in the tractogram that they represent. For example, when two tracts followed a similar path, then their corresponding point representations were placed in nearby points on the plane. Figure 3 shows a screenshot from a prototype tool using this representation.

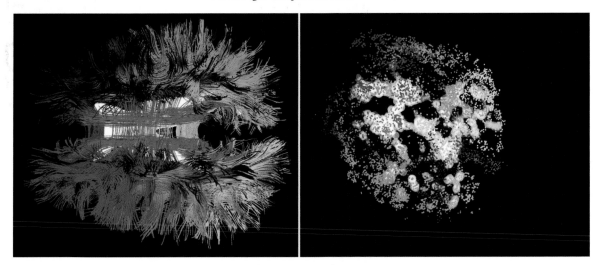

Figure 3 A 3D tractogram model (left) and its 2D point-set representation (right). Interactions on both representations are linked. For example, whenever a primitive (tract or point) is selected in either representation, then the selection is reflected in both. The current selection is higlighted with red in both views. Tracts (and their corresponding point representations) are colored based on their similarity: similar tracts are assigned similar colors and dissimilar tracts are displayed with dissimilar colors.

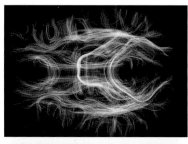

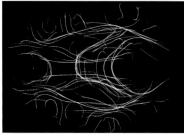

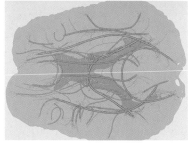

Figure 4 Iterations of the 2D neural path representation. In the final version (bottom), each stylized curve represents a bundle of tracts. Curve widths are modulated with the number of the tracts in the corresponding bundles. The blue region indicates the ventricles, a major anatomical landmark in the brain. The gray background is a horizontal silhouette of the brain surface. The final form of the representation was decided based on the degree to which it possessed the following characteristics: 1) simple but informative, consistent-multiscale axonal connectivity, 2) two-dimensional stylized look, and 3) useful anatomical context.

One common criticism we received from our neuroscientist collaborators, however, was that the 2D point-set representation did not have sufficient anatomical context by itself. In other words, it was too abstract for quick understanding.

Inspired partly by geographical maps and 2D illustrations seen in medical textbooks, our next approach was to represent neural tracts with 2D curves. This would also give the familiarity and anatomical context sought by the practitioners—remember neuroscientists are used to looking at cross-sectional images of the brain.

From the beginning we intended to take the notion of brain "mapping" literally by using the geographical map metaphor—producing a representation of the brain that can be viewed, interacted with, queried, and enriched just like an online geographical map. The highly popular online geographical map service Google Maps provided a practical way of implementing our idea on the web. We iterated several times while constructing the 2D path representation. Figure 4 shows a few intermediate results and the final representation.

Figure 5 shows screenshots from a standalone tool and a web application using the 2D neural path representation. Throughout this project, we had two common-sense strategies in mind to reduce visual complexity: abstraction and nested multiscale hierarchization. We believe that the resulting prototype was a successful implementation of these strategies. Our 2D path representation helps in fast multiscale exploration of axonal brain connectivity. Also, the web interface provides a familiar and efficient online interaction framework to share datasets, and, more importantly, it gives users the opportunity to integrate and enrich their data with the vast knowledge base of the web (see Figure 6).

INTERACTIVE GRAPHICS

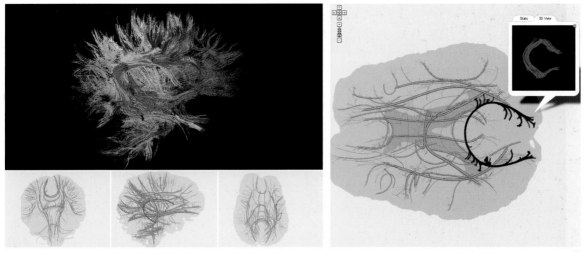

Figure 5 Schematic planar projections of DTI tractograms as part of a standalone interactive system (left) and as a web-accessible digital map (right). In the stand-alone application, interactions on the 3D and 2D views are linked. Shown in the pop-up window on the right is the "brain view" of the selected tract (analogous to "street view").

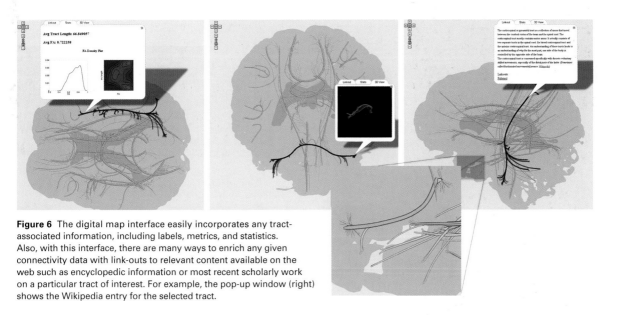

Figure 6 The digital map interface easily incorporates any tract-associated information, including labels, metrics, and statistics. Also, with this interface, there are many ways to enrich any given connectivity data with link-outs to relevant content available on the web such as encyclopedic information or most recent scholarly work on a particular tract of interest. For example, the pop-up window (right) shows the Wikipedia entry for the selected tract.

Delving into Image Features *by Eric Monson*

Eric Monson worked with Mauro Maggioni to create an interactive Graphical User Interface (GUI) to explore a new analytical method for large datasets developed by Maggioni's group at Duke. The analysis technique takes a large amount of high-dimensional data, such as images or text documents, and automatically detects features of the data at multiple levels of resolution for tasks such as data compression, classification, and outlier detection. The GUI has been successfully applied to multiple datasets analyzed with these methods; here they describe a project on the Yale Face Database B.

Over the years, people have used transformations to learn important characteristics of their data. Functions like sines, cosines, and wavelets can help describe the characteristics of low-dimensional data like audio signals and individual images, and be used as criteria to classify them, or to perform tasks such as compression. However, many important types of data are high-dimensional. For example, in a collection of images, each image can be said to have one dimension for each pixel in the image, so a group of 28×28 pixel images can be thought of as a cluster of points (one point for each image) in a 784-dimensional space. In this high-dimensional realm, there are no standard functions that apply to all the disparate data types. It is therefore necessary to learn "dictionaries" of characteristics directly from the data themselves. Mauro Maggioni's group at Duke has developed a technique called "Geometric Wavelets," which overcomes issues with previous techniques for this problem; it uses fast, intuitive algorithms and has guarantees on the quality of the interpretable results.

A sample of faces from the Yale Face Database B (5760 images, 10 subjects in 9 poses, each with 64 illumination conditions).

Geometric wavelets allow us to learn a "dictionary" of characteristic features of these data by breaking the data down into subsets. Within each subset, we find an "average" image, as well as "directions" and amounts in which images vary from the average. Then we break each subset into two and repeat the analysis.

INTERACTIVE GRAPHICS

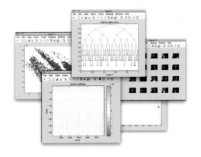

It would take hundreds of static figures to represent all of the analysis results. Only through an interactive Graphical User Interface (GUI) can we explore the data efficiently while retaining all of the links between the pieces.

When we started this project, Mauro and his postdoctoral associates were still developing the algorithms behind the method, but were looking ahead to when they would be explaining this technique to others and working with collaborators who wanted to apply their analysis to different types of data. The data analysis was being performed in Matlab, so they were generating a bunch of static plots and images. Some of these views gave a nice summary of fragments of the results, but if there are 2000 images being analyzed (which is still considered a "small" dataset), it isn't feasible to open up 2000 windows on the screen and try to sift through them and understand anything. In general, interactive exploration of large datasets should enable users to view more data, more quickly, while using a smaller amount of screen space. So we worked with the applied mathematicians to develop a graphical user interface for their datasets and analyses.

The first essential step was to meet with our collaborators to learn about the analysis technique and what data displays they wanted and needed to see. I took copious notes, trying to understand what each of their current figures was showing, and why, as well as the data structures I would be dealing with. This allowed me to implement their ideas and to suggest alternatives based on my experience in data visualization. Right away we also started to set priorities on which views were most important. Each additional plot might add more information, but a screen crowded with displays can dilute the message and make other items harder to see through distraction or size.

At this stage of a project, I strongly believe it's important not to rush to the computer too quickly. Pen and paper are wonderful tools for quickly playing around with ideas about layouts, and even interaction techniques, links between views, and internal data structures without being tied down to the constraints of computer illustration tools or programming languages.

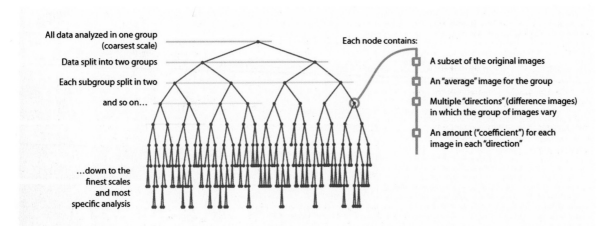

Our initial sketches of the layout. Even though they're messy, diagrams like these helped me clarify both the spatial layout and the internals of the program that I'd be writing. It's okay that some of it isn't pretty — you can generate nice figures based on your sketches when you want to communicate your ideas to others. We explored where to put different panels on the screen, and how to link the various views together.

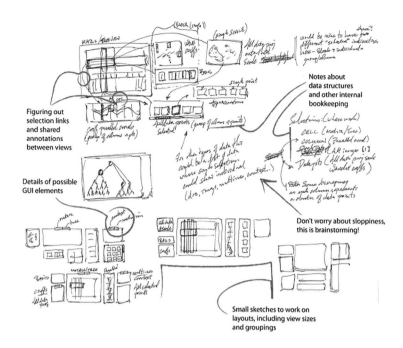

Figuring out selection links and shared annotations between views

Notes about data structures and other internal bookkeeping

Details of possible GUI elements

Don't worry about sloppiness, this is brainstorming!

Small sketches to work on layouts, including view sizes and groupings

Just as in a static data display or explanatory graphic, the layout of elements in an interactive visualization needs to effectively group common elements and keep visual clutter to a minimum, while maximizing the use of available space. We decided to present multiple views embedded in a single window. This is limiting in some ways, but with this arrangement you don't have to bring windows to the "front" before interacting with them, and you control what is seen and hidden. There were also constraints imposed by the natural shapes of certain views: scatterplots look better square, parallel coordinates with many axes are rectangular, etc. Other practical considerations forced certain decisions — for example, the visualization library I chose to use made it difficult to set the parallel coordinates plot vertically, so I had to adapt to a horizontal arrangement for that view.

The image data being analyzed in this example can be effectively displayed using grayscale values, but we decided to use color in two different ways. First, different types of selected elements have their own color, which clarifies the context of each element across multiple views. Color is also used to show quantitative variables in a diverging green-white-brown colormap for images that contain both positive and negative values. We tried to select colors that could be perceived and distinguished easily.

INTERACTIVE GRAPHICS

Images simultaneously show scatterplot axis "directions" and act as controls for X & Y axis choice to reduce visual clutter.

Direct navigation through the central tree view speeds exploration of the representation.

Selections of data subsets can be made in any view, and consistent colors show context across views.

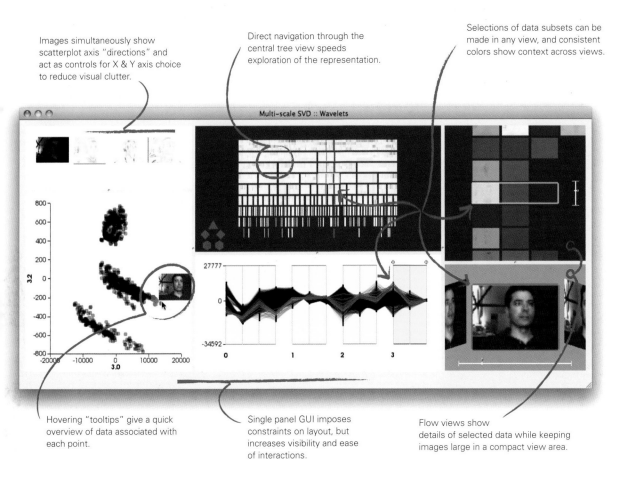

Multi-scale SVD :: Wavelets

Hovering "tooltips" give a quick overview of data associated with each point.

Single panel GUI imposes constraints on layout, but increases visibility and ease of interactions.

Flow views show details of selected data while keeping images large in a compact view area.

The final layout of the GUI. Users can select data from multiple panels, and other panels automatically update.

This is only the story of our first-generation GUI. As the project moves forward and methods behind the analysis change, we will have to make new decisions about how to keep the visualization effective and usable while incorporating additional views and adapting to new data types such as hyper-spectral images or text documents. Already, though, the ability of the mathematicians to see so much data so quickly has changed the way they are developing the underlying methods. Casually perusing the representation led to new insights into how to make their analyses more powerful as well as how the visualization should be modified as this process continues.

The Ribosome in Animation *by Said Sannuga*

The goal of this project was to produce an animation of the ribosome, a giant protein machine with many components that makes proteins from messenger RNA. The project was actually carried out in two stages. In the first stage, I produced an animation for myself, based on my research interests and background in molecular visualization technologies. Most of the information for this initial project came from a review written by Venki Ramakrishnan. Eventually, Dr. Ramakrishnan saw my animation and asked me to produce a more detailed version for him that would be appropriate for teaching undergraduates and for use in his seminar presentations. This discussion covers both stages.

When I set out to make an animation of the ribosome for myself, I decided that the animation should emphasize the structural and functional complexity of the ribosome. I wanted the visual style to be more biological looking, more organic, and more complex than conventional molecular visualizations. I did not make significant sketches at either stage of this project (apart from some technical sketches). Instead, I relied mostly on the images and illustrations available in the literature, as well as a large number of reviews and publications of the ribosome structure and functions.

To achieve a unique style for both projects, I did the following:
- Included all factors available with known structures.
- Minimized over-simplifications.
- Adjusted the speed, slowing some steps down so that complex interactions could be followed and speeding up some processes (to near real time) so that the viewer could appreciate the amazing speed at which these events take place.
- Used line representation in addition to surface representation; using both shows all the atoms involved and gives an overall sense of the structural complexity of the ribosome.
- Used morphing to show multiple states whenever two or more molecular structural states are known to demonstrate dynamic structural complexity.
- Decided not to include vibration, because while it would more closely represent the atomic behavior of molecules, it would distract from the main functions of the molecules involved.

INTERACTIVE GRAPHICS

Images from the first version of the animation

A specialized tRNA molecule charged with an amino acid (fmet-tRNA) binds to a messenger RNA, the small ribosomal subunit and initiation factors.

The two ribosomal subunits coalesce around the bound messenger RNA.

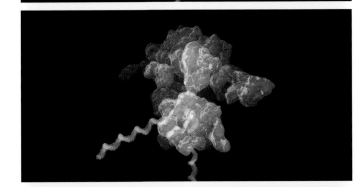

The next charged tRNA (pink) is carried to the complex by elongation factor Tu (yellow).

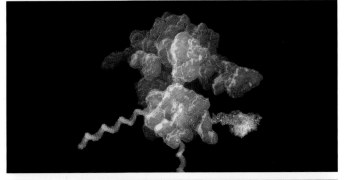

After the amino acid is added to the growing polypeptide chain, the tRNA is released and a new one binds.

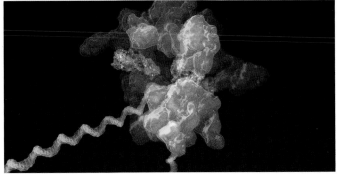

Producing this initial animation facilitated my eventual communication with Venki about how to increase accuracy and include more detailed information. In my original version, all subunits are presented in a unified color scheme to give a sense of the ribosome as one large machine. Venki advised me to color some components differently, so that they would be easier to refer to. For this version, we chose to color the two major subunits differently and to give distinct colors to many individual auxiliary proteins. The subtle use of lighting, shading, and image processing techniques also helped to differentiate the subunits. We also decided to include more detail about particular key steps — both by labeling components and steps, and by showing some specific atomic interactions at high resolution. After producing several storyboards and discussing these on several occasions with Venki, I produced the preliminary design of the project.

Images from the second version of the animation

The updated rendering of three tRNA molecules bound to the ribosome. Note the use of different colors to distinguish the ribosomal subunits and the tRNAs from one another.

tRNAs are shown entering and exiting the ribosome/mRNA complex.

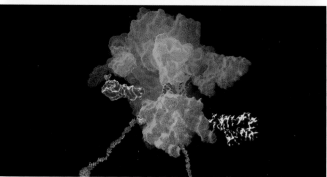

INTERACTIVE GRAPHICS

In this version of the animation, we often used labels to clarify particular steps and molecules. Here, each tRNA binding site and tRNA molecule is labeled.

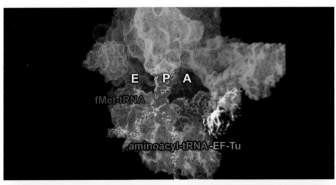

We also showed various tRNAs and auxiliary proteins without surface renderings to emphasize the chemistry of the peptidyl transfer reaction. We also simplified the representation of the amino acid to make the reaction easier to follow. Here the auxiliary protein EF-Tu is shown in yellow, conjugated to an amino acyl tRNA in blue.

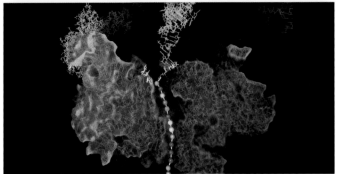

We labeled some steps in more detail—here the termination step is mediated by the auxiliary protein RF2 (shown in pink).

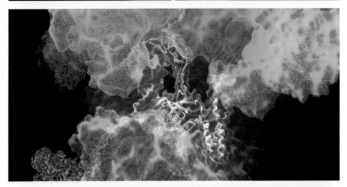

We also zoomed in to show some specific atomic interactions; here we show an interaction between the tRNA, the ribosome, and the mRNA in detail.

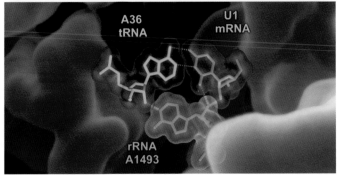

To execute this complex project, I had to develop a workflow to translate the illustrations and images into an accurate molecular animation based on real data. When I was working on this project, I was not aware of any significant molecular simulations or animations available that could guide me in the process. I carried out multiple steps including the following:

- Selecting the molecular structures (Protein Data Bank files containing the spatial coordinates of all atoms in each molecule).
- Identifying the software required and developing some scripts.
- Experimenting with various molecular representation styles (using molecular visualization and general purpose animation software) and identifying the most effective way of representing the different structures for this particular molecular system.

I prepared the structures in molecular visualization software and using some scripts that I developed. All subsequent modeling, objects management (e.g. grouping of the components), shading, lighting, animations, dynamics, and rendering were done using Maya (Autodesk). I used Fusion (Eyeon) to composite the Maya images.

After importing the structures into Maya, I first combined them into different groups and layers and animated them using classical animation tools or dynamic simulations. The next stage was to set up the shading (visual properties of the surfaces) and the lighting of the various components. This is closely tied to a person's particular viewpoint, rather than a fixed point in space. For example, some of the lights were tied to the viewpoint so that whenever the viewpoint changed, the lights maintained the same position. An analogy would be a miner's headlight—it's set up to shine light wherever you turn your head. Some of the shading attributes are also tied to the viewpoint, such as the "facing ratio," in which those surface areas that face the camera and those areas that face away from the camera can be assigned different properties. Using transparency, this would be like X-ray vision—whatever you look at becomes more transparent.

It is typical to decide on the exact shading and coloring scheme of all the components at this stage of a project (that is, at the pre-rendering stage) because rendering is the most time-consuming step and scientists who develop their own animations want to do it only once. However, I decided to use a strategy commonly used by professional animators, in which they render many individual layers and combine them at the compositing stage. This gives the animator more freedom to change visual attributes of the individual components. I chose to include surface attributes such as the lighting and the facing ratio for both the front and the back surfaces of the objects at this early stage so that I could experiment with these to get the right set of parameters.

INTERACTIVE GRAPHICS

To use this approach, I needed to know the exact depth (known as z-depth) of each surface element from the camera viewpoint, so that surfaces present in separate image layers can be recombined accurately at the compositing stage. This is a very common technique that can also be used to generate a color gradient (e.g., fog) to clarify the depth of objects in an image. These same rendering methods can also be used to give visual depth to the line elements (atomic representation style)—without them, the line elements look flat. There are other ways to represent the molecular backbone that have more dimensionality, such as the more common ball and stick representation. However, the ball-and-stick representations require a large amount of memory to handle a structure this large.

The two animations were actually produced from the same set of image layers with slight parameter variations—and you can see that they look quite different. Compositing many layers in this way might not be possible for the non-expert animator, but at the very least one should be aware that multiple lighting characteristics—including the position, intensity, color, and type (e.g., diffuse or ambient, etc.)—are extremely important for the visual clarity of an animation, especially if the viewpoint moves.

For scientists who are interested in using Maya software to produce molecular animation, there are currently many tutorials available on the web (such as www.molecularmovies.com). There are also a few tools available for managing molecular data inside Maya. The main tool is Molecular Maya (or mMaya for short) a plug-in which allows for the import and management of molecular data inside Maya in a powerful and user friendly manner.

The final products are two different image styles that aimed to achieve two different outcomes. The first style was meant to focus on the appreciation of complexity and to give the impression that the animation is of biological origin. The second, detailed animation had a similar purpose, but I also wanted it to be more communicative and informative by using more colors that could distinguish the different molecular components.

Animating the Reconstruction of a Silicon Surface *by Yan Liang*

Materials scientist Yan Liang used historical static images as a starting point to create an explanatory animation.

When a clean surface is produced by crystal cleavage, high-energy dangling bonds (or broken bonds) are generated at the surface. To minimize the number of the dangling bonds and hence the energy, surface atoms often rearrange themselves and form new bonds. The new atomic arrangement, which is different from that of the bulk crystal, is referred to as reconstruction.

Si(111)–7×7 reconstruction is a sophisticated and probably the most famous surface structure. Discovered in 1959, it was finally solved by K. Takayanagi et al. in 1985 (see Visual Index), concluding over 25 years of research activities of the surface science community. Even today, Si(111)–7×7 reconstruction still serves as a test bed for new surface analysis techniques due to its complexity.

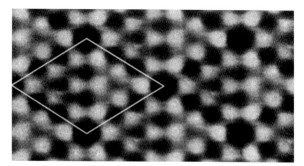 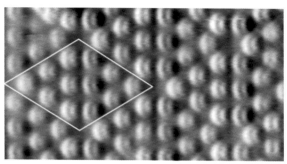

Left The very first Scanning Tunneling Microscope (STM) image of Si(111)–7×7 reconstruction, published by G. Binnig and H. Rohrer et al. in 1983 (see Visual Index). Binnig and Rohrer were awarded the Nobel Prize in 1986 for inventing the STM.

Right The Atomic Force Microscope (AFM) image revealing the subatomic features of Si(111)–7×7 reconstruction, published by Franz J. Giessibl et al. in 2000 (see Visual Index). The primitive unit cells are highlighted by the white outline. The major features seen in the images are the adatoms. Both images were cropped and resized to fit each other.

When I first started studying Si(111)–7×7 reconstruction with the aid of static diagrams (a famous diagram is shown on the top of the next page) and text in research papers, it took me hours to fully understand its structure and its relationship with the ideal unconstructed Si(111) surface. This motivated me to make an animation about this complicated structure. The main goal was that once a viewer finished watching the animation in a period of several mintues, he/she should have a good understanding of the structure of Si(111)–7×7 reconstruction.

To meet this goal, first I chose a new style to represent the surface. I added colors to distinguish atoms within different layers. I also added tetrahedrons to indicate that atoms are not on the same plane.

Second, I animated the transformation from an ideal unconstructed Si(111) surface to a constructed one through a series of imaginary steps. This way the relationship between the unconstructed and constructed surface becomes more apparent. Since reconstruction occurs only in the first bilayer, I decided to omit the rest of the layers in the animation, so that the viewer can focus on the transformation within the first bilayer.

⊕ | INTERACTIVE GRAPHICS

Diagram of Si(111)–7×7 reconstruction by K. Takayanagi et al. 1985 (see Visual Index), who solved the structure in 1985. Primitive unit cell is highlighted by the red outline.

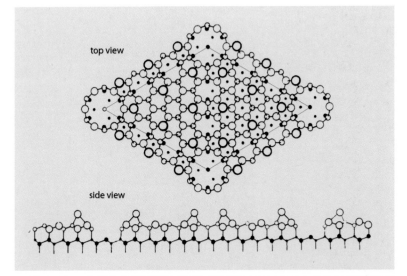

New diagram of the primitive unit cell of Si(111)–7×7 reconstruction. Colors and tetrahedrons were added to indicate atoms are not on the same plane.

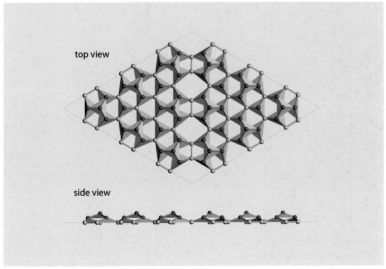

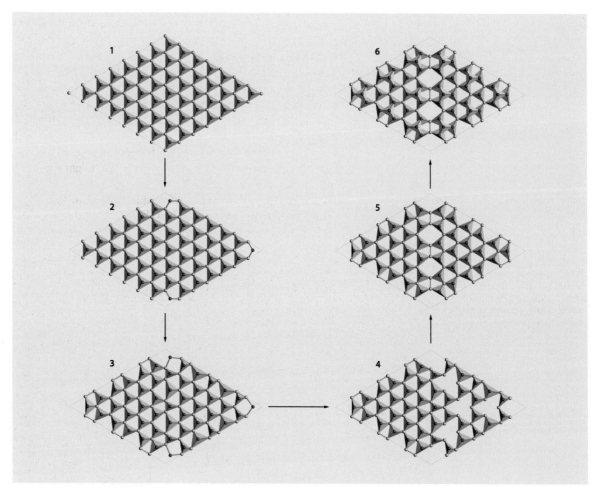

Imaginary steps transforming an unconstructed 7×7 structure to a primitive unit cell of Si(111)–7×7 reconstruction. The transformations between steps were animated.

Another important goal was to make the animation accessible to people who are not familiar with the terminology of surface science. I imagined they might have questions such as the following:

- What is a primitive unit cell? What does "7×7" mean?
- What is a Si(111) surface? How does it relate to the Si crystal?
- What is a "bilayer"?
- What is the structure of the Si crystal?

To provide answers in the animation, I started from a single Si atom and showed it growing into an Si crystal. Then, the crystal was cleaved to generate a Si(111) surface, and the concept of "bilayer" was defined. I then defined the primitive unit cell and 7×7 structure on the Si(111) surface. Next, I animated the transformation from an unconstructed 7×7 structure to a primitive unit cell of the constructed surface. Finally, by repeating the newly formed primitive unit cells, the reconstructed surface was generated.

INTERACTIVE GRAPHICS

I imagined the majority of the audience for this animation would be people who want to learn about the Si(111)−7×7 reconstruction. Consequently, I chose a clean and minimalistic visual style, and gave the animation a "motion diagram" look. The visual elements should help the audience to learn with minimal distraction:

- Only one perspective camera was used to show the Si crystal structure, all the other cameras are orthographic cameras to give accurate projections.
- A calm color palette was chosen to help the viewer to concentrate on the structures. Saturated red color was used to highlight special information or structure changes.
- Captions were kept as short as possible. Some captions appeared where they could serve their purpose best. In addition, using graphics directly inside the caption made it easier and faster to follow.

Two screenshots of the animation. Notice the color palette and the position of the captions in the top image, and the use of graphics inside the caption in the bottom image.

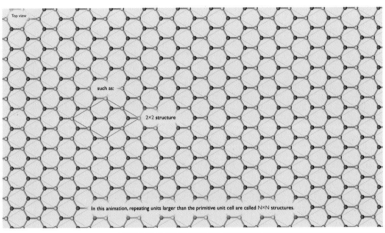

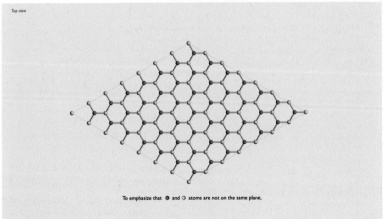

137

Exploring the Brain *by Drew Berry*

Drew Berry was asked to create an animation for the Whole Brain Catalog (WBC), part of the Whole Brain Project. The WBC is an open source, multiscale virtual catalog of the mouse brain that allows scientists and other users to integrate data in a common multiscale spatial framework. The WBC environment allows viewing of multiple datasets, such as 2D brain slices and tomograms, 3D representations of brain parts, EM volumes, 3D cell morphologies, and protein expression models. The Whole Brain Project is run by Dr. Mark Ellisman and his team at University of California, San Diego, and funded by the Waitt Family Foundation.

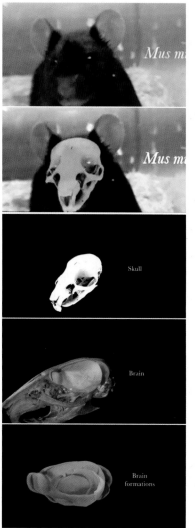

The purpose of this animation was to demonstrate the experience of exploring data on a fully operational Whole Brain Catalog. Within a timeframe of four months, I had to integrate many forms of raw neuroscience data that were derived from different magnitudes of both size and timescales: from a live mouse, down through brain regions and tissues, to a single neuron (snapshots at left). The objective was to allow the viewer to travel over the surface of the neuron to see its internal organelles and other structures before arriving at the location where a new synapse would form. Timescales would have to be seamlessly manipulated to make many phenomena viewable to an audience, including real-time live mouse footage, neuron tissue 40Hz firing patterns, and synapse formation that can only be observed through time-lapse microscopy.

3D models were available for the larger structures, such as a computed tomography (CT) mouse skull and gross brain formations, but smaller features had to be reconstructed from a composite of many forms of data. One of the biggest challenges was constructing a single tissue layer of neuron cells in the *Dentate gyrus* that lies within the hippocampal formation. Through resources such as BrainMaps.org, I had access to 2D serial slices of the mouse brain that were stained to highlight the cell types that I was interested in. With these data I was able to step through the brain one slice at a time and build a 3D outline of where these particular neurons were in the *Dentate gyrus*. I used this outline model to construct simulated granule neurons that were positioned and oriented correctly, and displayed accurate representations of their axon and dendrite structure.

Most of the data I used for this visualization were obtained from web resources (ncmir.ucsd.edu, ccdb.ucsd.edu/index.shtm and BrainMaps.org). On the following pages are example data and models that I derived from those data.

INTERACTIVE GRAPHICS

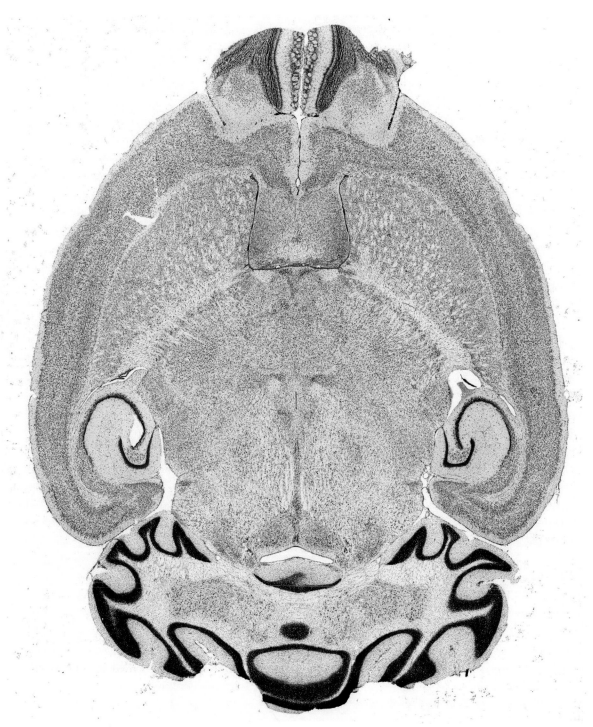

A horizontal slice through an adult mouse brain, Nissl stained to highlight the cell bodies of the neurons from BrainMaps.org.

A higher-resolution view of the slice through the right hippocampal region, with the dense layer of granule neurons forming a dark "U" in the middle of the image.

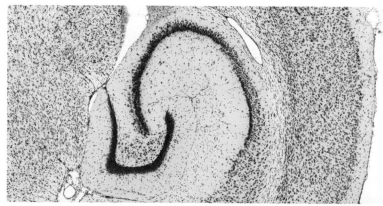

A reconstruction of the shape of the *Dentate gyrus* made by stepping through the brain slices and manually drawing curves that follow the shape of the granule neuron cell layer. I then connected these curves in 3D space to create a model of this tissue layer. This model was populated with simulated granule neuron cells that can be seen in the subsequent image.

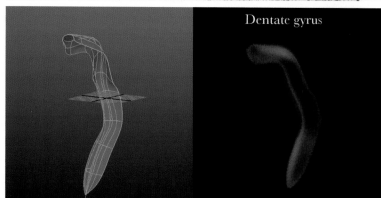

Dentate gyrus

As the camera zooms from the tissue scale down to the cellular scale, a single 2D slice of the *Dentate gyrus* is revealed; it's populated with simulated neuron cells. The collection of granule neurons is presented as living tissue and exhibit rhythmic collective firing which references the 40Hz "gamma oscillation" firing pattern that can be observed in these cells in vivo. A 40Hz cycle would not be perceptible to the audience—it's too fast—so I took the liberty of slowing the activity to about 0.5Hz or so.

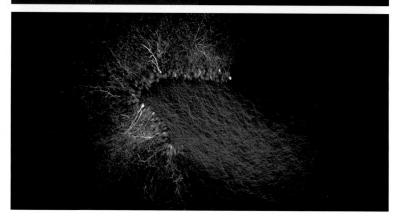

INTERACTIVE GRAPHICS

I was able to accurately model individual neurons with cell body (soma), dendrites, and axons drawn to scale. The cell's axon, which is heading off to the right in this image, was reconstructed from multiple data types.

This is an example of the data used to reconstruct individual neurons. This particular image is a tomographic reconstruction of a mouse neuron dendrite branch.

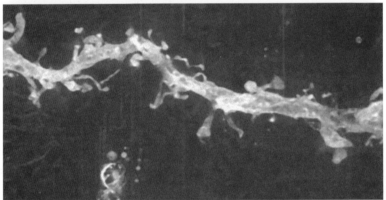

As you zoom in even further, the cell body (soma) and its internal organelles are also accurately represented. These models were determined from tomography reconstructions.

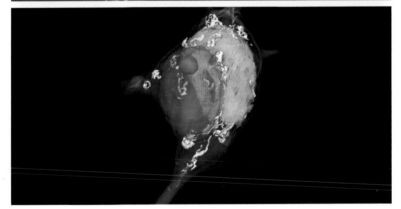

VISUAL
INDEX

16

17

18

19

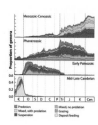

20

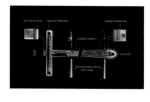

21

22

23

24

25

16 Nichols, B. (2005). Cell biology–Without a raft. *Nature,* 436(7051): 638–639.

17 Katul, G., Porporato, A., & Oren, R. (2007). Stochastic dynamics of plant-water interactions. *Annual Review of Ecology Evolution and Systematics,* 38: 767–791.

18 Seeley, T. D., & Buhrman, S. C. (1999). Group decision making in swarms of honey bees. *Behavioral Ecology and Sociobiology,* 45(1): 19–31.

19 Yin, P. (2010). Adapted from a slide presentation; Wyss Institute and Department of Systems Biology, Harvard Medical School.

20 Thirumalai, D., Lee, N., Woodson, S. A., & Klimov, D. K. (2001). Early events in RNA folding. *Annual Review of Physical Chemistry,* 52: 751–762.

21 Fowlkes, C. and DePace, A. (2010). Adapted from a slide presentation. Department of Systems Biology, Harvard Medical School.

22 Foster, J. Full cloud data for Perseus, Ophiuchus and Serpens. Retrieved 10 July 2011, from www.worldwidetelescope.org/COMPLETE/WWTCoverageTool.htm

23 Bush, A. M., & Bambach, R. K. (2010). Paleoecologic megatrends in marine metazoa. *Annual Review of Earth and Planetary Sciences,* 39(1): 241–269.

24 Losey, M. W., Schmidt, M. A., & Jensen, K. F. (2001). Microfabricated multiphase packed-bed reactors: Characterization of mass transfer and reactions. *Industrial & Engineering Chemistry Research,* 40(12): 2555–2562.

25 Dabbousi, B. O., RodriguezViejo, J., Mikulec, F. V., Heine, J. R., Mattoussi, H., Ober, R., Jensen, K. F., & Bawendi, M. G. (1997). (CdSe)ZnS core-shell quantum dots: Synthesis and characterization of a size series of highly luminescent nanocrystallites. *Journal of Physical Chemistry B,* 101(46): 9463–9475.

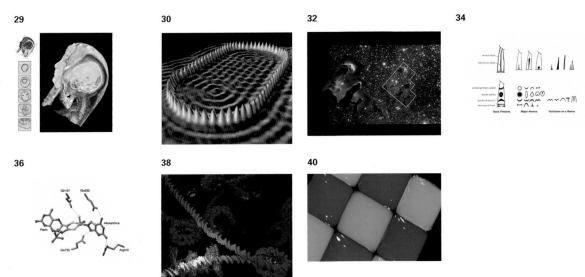

29

30

32

34

36

38

40

29 Frankel, F. (2006). Slicing into the past. *American Scientist,* 94: 550–551.

30 Crommie, M. F., Lutz, C. P., & Eigler, D. M. (1993). Confinement of electrons to quantum corrals on a metal surface. *Science,* 262(5131): 218–220.

Caption quoted from *American Scientist,* Sightings, 93, 2005 May–June.

32 Pasachoff, J. M. (2002). *Astronomy, from the Earth to the Universe* (6th ed.). Brooks/Cole–Thomson Learning, Pacific Grove, CA.

Images produced by P. Scowen, J. Hester, Mark McCaughrean, and M. Andersen.

34 Nijhout, H. F. (1991). *The development and evolution of butterfly wing patterns*. Smithsonian Institution Scholarly Press, Washington, DC.

36 Pauff, J. M., Cao, H., & Hille, R. (2009). Substrate orientation and catalysis at the molybdenum site in xanthine oxidase crystal structures in complex with xanthine and lumazine. *Journal of Biological Chemistry,* 284(13): 8751–8758.

38 McGill, Gaël. (2007). Unpublished. Digizyme, Inc.

40 Abbott, N. L., Folkers, J. P., & Whitesides, G. M. (1992). Manipulation of the wettability of surfaces on the 0.1-micrometer to 1-micrometer scale through micromachining and molecular self-assembly. *Science,* 257(5075): 1380–1382.

45

46

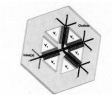

48

50

52

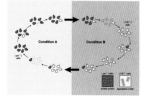

54

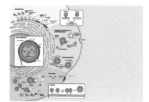

56

45 Meitl, M. A., Zhu, Z. T., Kumar, V., Lee, K. J., Feng, X., Huang, Y. Y., Adesida, I., Nuzzo, R. G., & Rogers, J. A. (2006). Transfer printing by kinetic control of adhesion to an elastomeric stamp. *Nature Materials,* 5(1): 33–38.

46 Albrecht, J., & Jensen, K. (2006). Presentation, Department of Chemical Engineering, MIT.

48 Garstecki, P., Gitlin, I., DiLuzio, W., & Whitesides, G. M. (2004). Formation of monodisperse bubbles in a microfluidic flow-focusing device. *Applied Physics Letters,* 85(13): 2649–2651.

50 Orengo, C. A., & Thornton, J. M. (2005). Protein families and their evolution—A structural perspective. *Annual Review of Biochemistry,* 74: 867–900.

52 Revised image based on early sketches for Halfmann, R., & Lindquist, S. (2010). Epigenetics in the extreme: Prions and the inheritance of environmentally acquired traits. *Science,* 330(6004): 629–632.

54 Krahmer, N., Guo, Y., Farese, R. V., & Walther, T. C. (2009). SnapShot: Lipid droplets. *Cell,* 139(5): 1024–1192.

56 Helmuth, B., Mieszkowska, N., Moore, P., & Hawkins, S. J. (2006). Living on the edge of two changing worlds: Forecasting the responses of rocky intertidal ecosystems to climate change. *Annual Review of Ecology Evolution and Systematics,* 37: 373–404.

VISUAL INDEX

61

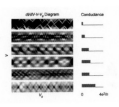

62

64

66

68

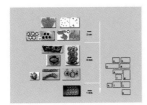

70

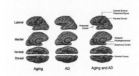

72

61 Werner, T., Koshikawa, S., Williams, T. M., & Carroll, S. B. (2010). Generation of a novel wing colour pattern by the Wingless morphogen. *Nature,* 464(7292): 1143–1157.

62 Carol, R. J., Takeda, S., Linstead, P., Durrant, M. C., Kakesova, H., Derbyshire, P., Drea, S., Zarsky, V., & Dolan, L. (2005). A RhoGDP dissociation inhibitor spatially regulates growth in root hair cells. *Nature,* 438(7070): 1013–1016.

64 Cant, R. S., Dawes, W. N., & Savill, A. M. (2004). Advanced CFD and modeling of accidental explosions. *Annual Review of Fluid Mechanics,* 36: 97–119.

66 Kroes, G. J. (2008). Frontiers in surface scattering simulations. *Science,* 321(5890): 794–797.

68 Liang, W. J., Bockrath, M., & Park, H. (2005). Transport spectroscopy of chemical nanostructures: The case of metallic single-walled carbon nanotubes. *Annual Review of Physical Chemistry,* 56: 475–490.

70 Whitesides, G. M. (2002). Unpublished. Department of Chemistry and Chemical Biology, Harvard University.

72 Buckner, R., Salat, D., Dickerson, B., Rosen, B., Athinoula, A. (2010). Adapted from slide presentation. Martinos Center for Biomedical Imaging, Department of Radiology, Massachusetts General Hospital.

77

78

82

84

88

94

96

102

77, 102 Stephens, A. D., Haase, J., Vicci, L., Taylor, R. M., & Bloom, K. (2011). Cohesin, condensin, and the intramolecular centromere loop together generate the mitotic chromatin spring. *Journal of Cell Biology,* 193(7): 1167–1180.

78 Venn, J. (1880). On the diagrammatic and mechanical representation of propositions and reasonings. *Philosophical Magazine and Journal of Science,* S.5 9(59).

Baron, M. (1969). A note on the historical development of logic diagrams: Leibniz, Euler and Venn. *Mathematical Gazette,* 53:113–125.

Figure 3B, Redrawn from Robinson, M. D., and Speed, T. P. (2007). A comparison of affymetrix gene expression arrays. *BMC Bioinformatics,* 8: 449.

82 Bryan, A. W. J., Starner-Kreinbrink, J. L., Hosur, R., Clark, P. L., & Berger, B. (2011). Structure-based prediction reveals capping motifs that inhibit β-helix aggregation. *PNAS,* 108(27): 11099–11104.

84 Rowat, A. C., Birda, J. C., Agrestia, J. J., Randob, O. J., & Weitz, D. A. (2009). Tracking lineages of single cells in lines using a microfluidic device. *PNAS,* 106(43): 18149–18154.

88 Massey, R. J., Christensen, L. L., & Frankel, F. (2007). Dark matter comes to light. *American Scientist,* 95: 257–259.

94 Young, A. Astrophysics: Cosmic jet engines. *Nature,* 463(7283): 886–887.

Meier, D. L. et al. (2001). Magneto-hydrodynamic production of relativistic jets. *Science,* 291(5501): 84–92.

96 Goodman, A., Udomprasert, P., & Sadler, P. (2010). Unpublished.

INTERACTIVE GRAPHICS

111

112

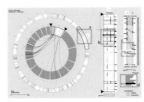

116

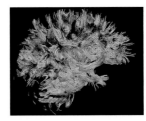

120

124

128

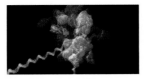

134

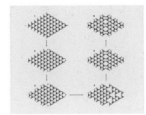

138

111 Davis, T. L., Walker, J. R., Campagna-Slater, V., Finerty, P. J. Jr, Paramanathan, R., et al. (2010). Structural and biochemical characterization of the human cyclophilin family of peptidyl-prolyl isomerases. *PLoS Biol,* 8(7): e1000439.

112 Manolio, T. A. (2010). Genomewide association studies and assessment of the risk of disease. *NEJM,* 363(2): 166–176.

116 Ma, L-J., et al. (2009). Genomic analysis of the basal lineage fungus Rhizopus oryzae reveals a whole-genome duplication. *PLoS Genetics,* 5(7): e1000549.

Meyer, M., Munzer, T., & Pfister, H. (2009). MizBee: A multiscale synteny browser. *IEEE Trans Vis Comput Graph,* 15(6): 897–904.

120 Jianu, R. Demiralp, C., & Laidlaw, D. H. (2011). Exploring brain connectivity with two-dimensional neural maps. *IEEE Transactions on Visualization and Computer Graphics,* 99.

124 Monson, E. E., Chen, G., Brady, R., & Maggioni, M. (2010). Data representation and exploration with geometric wavelets. Paper presented at the IEEE VAST Symposium.

128 www.molecularmovies.com/movies/viewanimatorstudio/Said%20Sannuga/

134 Liang, Y. (2008, 10 July 2011). Si(111) surface 7x7 reconstruction. Animation at vimeo.com/1086112.

Tanishiro, Y., Takahashi, S., & Takahashi, M. (1985). Structure analysis of Si(111)-7x7 reconstructed surface by transmission electron diffraction. *Surface Science,* 164(2-3): 367–392.

Binnig, G., Rohrer, H., Gerber, C. H., & Weibel, E. (1983). 7x7 reconstruction on Si(111) resolved in real space. *Physical Review Letters,* 50(2): 120–123.

Giessibl, F. J., Hembacher, S., & Bielefeldt, H. (2000). Subatomic features on the silicon (111)–(7x7) surface observed by atomic force microscopy. *Science,* 289(5478): 422–425.

138 ncmir.ucsd.edu/; ccdb.ucsd.edu/index.shtm; and BrainMaps.org

APPENDIX

APPENDIX

VISUAL STRATEGIES ONLINE FORUM AND MAGAZINE

We are pleased to offer our readers a "live" extension of Visual Strategies *in a magazine and discussion format at www.visual-strategies.org.*

Forum

In the Forum section you will have the opportunity to discuss a number of issues that lie beyond the scope of a printed book with colleagues, expert graphic designers, and journal editors. We welcome your opinions about the examples in the guide and discussion of your own challenges with regard to posters, journal covers, and so forth.

What's Wrong with This Picture?

In this section, we will post figures or interactive graphics for the purpose of discussion around the principles described in this guide. We see this as a means of revealing different approaches to graphical thinking.

Visual Strategies Links

Here you will find direct links to our many generous contributors and to the animations and interactive graphics examples from the guide. Readers will agree, we are certain, that doing justice to interactive graphics and animations is impossible in a printed format.

Education

In this section, we will provide a place for discussions and links to programs addressing the potential of creating graphics as a teaching and evaluation tool. The NSF-funded program, Picturing to Learn, www.picturingtolearn.org, developed the concept that creating visual representations of science and engineering becomes a powerful tool for discovering misconceptions among undergraduate students. Our visual strategies online magazine and web forum will connect to that already-ongoing conversation and encourage an online education forum for teachers at all levels of education.

Other Links

This section will connect to interesting and relevant design and science sites addressing representation.

FURTHER READING

Abbott, Berenice. *Photographs*. New York: Horizon Press, 1970.

Abbott, Edwin A. *Flatland*. 1884. Reprint, New York: Penguin Books, 1998.

Bringhurst, Robert. *The Elements of Typographical Style, Second Edition*. Point Robbers, WA: Harley & Marks Publishers, 2002.

Carter, David A. *One Red Dot*. New York, NY: Little Simon, 2004.

Cottin, Menena, and Rosana Faria. *The Black Book of Colors*, Fifth Printing. Toronto: Groundwood Books, 2009.

Edgerton, Harold. *Stopping Time, The Photographs of Harold Edgerton*. New York: Harry N. Abrams Publishers, Inc., 1987.

Frankel, Felice. *Envisioning Science, the Design and Craft of the Science Image*. Cambridge, MA: MIT Press, 2002.

Frankel, Felice. 2003, Jul-Aug - 2007, Jul-Aug. Sightings. *American Scientist*, 91(4)-95(4).

Frankel, Felice C., and George M. Whitesides. *No Small Matter, Science on the Nanoscale*. Cambridge, MA: Harvard University Press, 2009.

Frankel, Felice, and George M. Whitesides. *On the Surface of Things: Images of the Extraordinary in Science*. Cambridge, MA: Harvard University Press, 2009.

Friedman, Mildred. *Graphic Design in America: A Visual Language History*. New York: Harry N. Abrams, Inc., 1989.

Helfand, Jessica. *Reinventing the Wheel*. New York: Princeton Architectural Press, 2002.

Hooke, Robert. *Micrographia*. London, 1665. Reprint, Science Heritage Ltd., 1987.

Lupton, Ellen. *Thinking with Type: A Critical Guide for Designers, Writers, Editors, & Students*. New York: Princeton Architectural Press, 2010.

Mitchell, William J. *The Reconfigured Eye: Visual Truth in the Post-Photographic Era*. Cambridge, MA: MIT Press, 1992.

Morrison, Philip and Phylis, and the Office of Charles and Ray Eames. *Powers of Ten, About the Relative Size of Things in the Universe*. New York: Scientific American Library, 1982.

Müller-Brockmann, Josef. *Grid Systems in Graphic Design (original title Raster systeme für die visuelle Gestaltung)*. Sulgen, Switzerland: Verlag Niggli, 1996.

Pauling, Linus, and Roger Hayward. *The Architecture of Molecules*. San Francisco: W. H. Freeman and Co., 1964.

Rand, Paul. *Paul Rand: A Designer's Art*. New Haven, CT: Yale University Press, 1985.

Reich, Hanns. *The World from Above*. New York: Hill and Wang, 1966.

Rossiti, Hazel. *Colour: Why the World Isn't Grey*. Princeton, NJ: Princeton University Press, 1983.

Sagmeister, Stefan. *Made You Look*. New York: Abrams, 2009.

Sagmeister, Stefan. *Things I Have Learned in My Life So Far*. New York: Abrams, 2008.

Stevens, Peter S. *Patterns in Nature*. Boston: Atlantic Monthly Press, 1974.

Spiekermann, Eric, and E. M. Ginger. *Stop Stealing Sheep and Find Out How Type Works*. Mountain View, CA: Adobe Press, 1993.

Time-Life Series: *Life Library of Photography*. Alexandria, VA: Time-Life Books, 1971 (out of print).

Tschichold, Jan. *Asymmetric Typography*. Translated by Ruauri McLean. New York: Reinhold Publishing, 1967.